THE BELL

Circles of Awareness: The Metaphysical Power of Psychic Control

Darno Von DeJohnette

DeJohnette Publishing Group

Copyright © 2018, 2025 Darno Von DeJohnette

All rights reserved

The characters and events portrayed in this book are fictitious. Any similarity to real persons, living or dead, is coincidental and not intended by the author.

No part of this book may be reproduced, stored in a retrieval system, or transmitted in any form or by any means, electronic, mechanical, photocopying, recording, or otherwise, without the express written permission of the publisher.

First Edition: January 2018
Second Edition (Revised and Updated): December 2025

The Bell / Darno DeJohnette. -- 2nd ed.

ISBN-13: 9798744198374 (Paperback)

Cover design by: Darno DeJohnetter
Library of Congress Control Number: 2018675309
Printed in the United States of America

DEDICATION

To the Ancestors who have gone on before us, your lives, your struggles, and your triumphs have carved the path on which we walk today. Your legacy is the foundation of our existence, and your spirit guides us as we navigate the mysteries of life and the universe. We honor your memory and strive to build upon the wisdom you have passed down through the ages.

To the Mystics who brought us knowledge and granted us the wisdom to use it wisely, your insights and revelations have illuminated the darkest corners of our understanding. You have bridged the gap between the seen and the unseen, the known and the unknown, offering us a glimpse into the profound truths that lie beyond the veil of ordinary perception. Your teachings inspire and guide us, urging us to seek more profound meaning and greater harmony in our lives and the cosmos.

To the Theoretical Physicists who have embraced the concepts of the "multi-verse" and the relationship between quantum mechanics and human consciousness, Your bold inquiries and groundbreaking theories have expanded the horizons of human knowledge. You have ventured into the frontier of the very small and vast, revealing a far more intricate and interconnected reality than we ever imagined. Through your work, you have shown us that the boundaries between science and spirituality are illusions and that consciousness itself may be the key to unlocking the universe's secrets.

This dedication is a tribute to your collective contributions, courage to explore the unknown, and enduring quest for truth. May we continue to build upon your legacy with curiosity, humility, and a deep respect for the profound interconnectedness of all things.

"Metaphysics involves intuitive knowledge of unprovable starting-point concepts and truth, and demonstrative knowledge of what follows from them."

~ ARISTOTLE

CONTENTS

Title Page
Copyright
Dedication
Epigraph
Preface
Introduction
Prologue
Chapter One — 1
Chapter Two — 28
Chapter Three — 61
Chapter Four — 76
Chapter Five — 96
Chapter Six — 115
Chapter Seven — 135
Chapter Eight — 153
Chapter Nine — 171
Epilogue — 187
Books By This Author — 191
About The Author — 201
Acknowledgement — 203
AUTHOR CONTACT — 205

PREFACE

The Purpose And Essence Of The Books

The Books are presented as an atonement for my personal sacrifice. They are intended to reveal the profound nature of conscious awareness as the creative tool of God, the architect behind the master plan of existence. By building upon a foundation of expanding mental concepts, these works strive to unravel the mysteries of time and help us perceive truth within the reality of the present moment. Through the practice of God-consciousness, we become aware of a deeper state of being—one that enlightens our understanding and broadens our perspective. As awareness grows, we learn that there are many levels of fact, and we expand within the scope of greater understanding. God-consciousness instructs us, allowing us to recognize our thoughts interacting with a robust comprehension. Through this powerful awareness, we begin to see ourselves as conscious entities, capable of growth and spiritual development.

The Journey of Creation and Sacrifice

The Books were conceived under unique circumstances and serve as a reminder of the sacrifice made by distancing oneself from physical reality. In the process of creating these works, I devoted many of my youthful years to solitude, reflecting on the infinity of God and His magnificent creation. From August 1972 to September 1977, I was bound by an oath to seek and document a pathway into my subconscious mind. During this period, others observed as I spent countless nights and days writing my first manuscript—a testament to the power of dedication and the willingness to sacrifice in the pursuit of knowledge.

A Personal Transformation

As a result of this journey, I became something of an enigma to my family and friends. Despite being unable to control the powerful force that propelled my consciousness forward, I persisted in recording my experiences to the best of my ability. Words often felt inadequate to express the magnitude of the concepts I encountered, but I continued with the hope that, one day, my words might carry significance. Like many seekers before me—those who looked beyond the accepted theories and teachings of their time in search of deeper understanding—I made a choice. This was my sacrifice: a conscious decision to pursue knowledge and truth, a choice that empowers each of us to shape our own spiritual journey.

A Call to Seekers

As a seeker of knowledge, you must choose your own path to follow.

INTRODUCTION

Philosophical Perspectives

A philosophical document can aim for precision and still remain personal. In fact, the most useful philosophy often begins where personal life becomes difficult to interpret—when ordinary moral vocabulary no longer fits the complexity of lived experience. People say, "right choice" and "wrong choice," "success" and "failure," as if life were a test with a visible answer key. Yet the world repeatedly demonstrates a harsher lesson: **a good decision can lead to a terrible outcome, and a terrible decision can lead to a good outcome.** Most people already know this intellectually. The difficulty is accepting it emotionally and ethically.

We want the world to be arranged so that outcomes correspond to desert. We want vigilance to prevent catastrophe and virtue to guarantee protection. When reality refuses these wishes, we reach for explanations that restore the illusion of moral accounting: blaming victims, glorifying winners, rewriting stories after the fact, or insisting that unseen forces "must have had a reason." Such explanations can provide comfort, but they often do so by distorting truth—and by smuggling cruelty into our judgments. The harmed become "reckless." The fortunate become "worthy." The dead become cautionary tales instead of human beings.

This essay does not try to solve the universe. It tries to help the reader become less haunted by it. To do that, it introduces three lenses. First, **yin–yang** offers a way of seeing reality as interdependence and transformation rather than isolated causes. It emphasizes that forces interact, exchange roles, and shift over time; balance is not a permanent achievement but an ongoing adjustment. Second, **the journey metaphor** reminds us that a life is not a single outcome but an unfolding path under uncertainty. A person cannot evaluate a road by its final mile alone. Third, **god-consciousness** is treated as a practice of awareness—an inner orientation that can improve decision-making without pretending to control fate. In this usage, it is not a guarantee of protection. It is a commitment to act from deeper values rather than reactive ego.

The reader should not expect an academic treatise. The aim is not exhaustive historical scholarship. The aim is coherence: to articulate a view of agency that neither inflates the power of choice nor diminishes its moral significance. The moral practice recommended throughout can be stated plainly: **judge yourself and others by what could reasonably be known, intended, and chosen at the time—while still taking consequences seriously as realities to respond to, repair, and learn from.** That ethic does not offer the narcotic of certainty. It does offer something better: a humane framework that leaves room for both responsibility and compassion, both grief and meaning.

PROLOGUE

Cognative Science

There is a kind of darkness that does not feel like darkness. It feels like productivity. It feels like normal life proceeding as it always has: the schedule, the appetite, the ache to be approved of, the subtle panic of wasted time, the private bargains made with the future. Most people never describe themselves as asleep. They describe themselves as busy.

Psychic slumber, as the older language calls it, is not a lack of intelligence. It is intelligence narrowed into a tube, turned toward survival and social signaling until the wider field of awareness falls out of frame. In that condition the mind can function brilliantly—earn money, form relationships, build devices, argue politics—while never recognizing that it is living inside a curated slice of existence.

The slumber is not caused by evil. It is caused by habit. The mind builds a world out of what it repeatedly attends to. What it does not attend to becomes unreal. Then, because it is unreal, the mind does not attend to it. The loop closes.

In this loop, the most reliable illusion is that the current map is the territory itself. The person believes the world is what appears in the beam of attention. The beam may be bright, but it is narrow.

This book is an instrument for widening that beam. It presents a ladder of ideas and a manual of practices. It is esoteric in its presentation because the inner life is encountered as symbol, dream, ritual, and direct knowing. It is science-annotated because the modern reader deserves intellectual honesty: not every inner claim can be proved, and not every profound experience is a passport to metaphysical certainty.

Yet the nightly doorway remains. Every night the mind enters a theater it did not build consciously. Every night it becomes convinced by scenes made of memory, emotion, and association. Every night it forgets the waking self. This is an empirical fact of ordinary human life, regardless of metaphysical belief.

And because the doorway is reliable, training is possible. The question is not whether everyone will become a mystic. The question is whether anyone will refuse to remain a sleepwalker.

Science Note: Contemporary cognitive science describes perception and meaning as active constructions shaped by attention, prediction, and memory. The "narrow beam" metaphor aligns with research on attentional selection and salience-driven capture. The book's goal of "widening awareness" parallels interventions that increase metacognition and attentional flexibility, though the metaphysical framing exceeds what science can confirm.

CHAPTER ONE

The Infinity of the Egg

A Philosophical Reflection on Metaphysical Symbolism, Ancestry, and the Boundless Cosmos.

I. Seeking the Truth of the EGG

Reaching into the unexplored regions of the mind, one finds a landscape few traverse—territories reserved for those who care deeply about the truth of reality and the essence of the egg. The egg emerges as a symbol, mysterious and profound, inviting seekers to pursue wisdom and guidance from ancestors whose voices echo through the depths of time. This journey is not only intellectual but spiritual, a quest for understanding that bridges the seen and unseen, the living and the departed.

In these uncharted domains of consciousness, the layers of existence unfold, revealing truths that lie beyond the surface. The egg stands as a metaphor for creation and the cyclical nature of life, holding within its shell the mysteries of genesis and transformation. As we delve into this metaphysical journey, we invoke the spirits of our ancestors, their wisdom a beacon to illuminate our path. Though they have passed into the depths of time, their essence remains, nurturing our pursuit of knowledge and enlightenment. By embracing this legacy, we connect with a universal consciousness, transcending the boundaries of time and space in a profound communion with collective wisdom.

II. Inherited Wisdom

Somewhere within the genetic code of our minds lies the knowledge and wisdom of those who came before us. Each generation refines and enhances this encrypted wisdom, allowing us to unfold the nature of our own existence. In the silent depths of our being, the legacy of our forebears rests dormant, encoded within the intricate strands of DNA—a repository not just of physical traits, but of collective experience and understanding.

As we evolve, we unlock this ancient reservoir, accessing insights that transcend individual lifetimes. Each generation adds a new layer of wisdom, building upon the foundations of the past. This continual process enriches our ability to navigate the complexities of existence, imparting a deeper sense of purpose and awareness. Through the unfolding of inherited wisdom, we come to recognize our place in the universe and the interconnected web of all life. Honoring this knowledge, we contribute to the ongoing evolution of human consciousness.

III. The Universe as an Egg

The universe, in its vast and unknowable infinity, can be seen as a capsule—an egg containing an infinite number of reflections within its structure. These are eggs of a lesser nature, yet each possesses its own infinity. Within these nested realities, the boundaries are fluid, each containing the seeds of countless others, all interconnected in a grand cosmic dance. The fractal nature of existence suggests that every part of the universe, no matter how small, harbors infinite complexity and depth.

By perceiving the universe as a series of nested infinities, we gain insight into the profound interconnectedness of all things. Every entity, regardless of its apparent significance, mirrors the greater whole, contributing to the boundless tapestry of existence. This perspective invites us to explore the infinite both within and around us, recognizing the extraordinary in the ordinary, and the vast in the minute.

IV. The Boundless Cosmos

Beyond the realm of the known universe, one can only imagine equivalent infinities encompassed by ever greater eggs. Infinity, in this sense, is conceptually inexhaustible —a multi-dimensional expanse that diminishes toward the microcosm of the egg while simultaneously expanding toward the macrocosm. In our quest to comprehend existence, we encounter a succession of nested realities, each revealing that infinity is not simply vast but endlessly unfolding.

This vision transforms our understanding of the universe into a dynamic interplay of scales and dimensions, where the very small mirrors the very large in endless recursion. The egg, as metaphor, encapsulates this dual nature, representing both the seed of potential and the boundless expanse of realized infinity. Embracing this theory, we acknowledge the limitless nature of reality and recognize the universe as an ever-evolving tapestry of interconnected infinities.

V. The Mystery of Existence

The true position of our existence within the grand scheme remains an enigma. Our reality is intricately woven into the fabric of this conceptual seed—the egg—held in balance by the magnetism of our conscious awareness. This enigmatic infinity binds us to the very nature of the egg, inviting us to contemplate our place within the cosmic order.

Our consciousness acts as a focal point, a magnetic force interacting with surrounding infinity. This dynamic relationship suggests that our awareness is not merely a passive observer but an active participant in the unfolding of reality. The egg, as symbol, encapsulates all possibilities, binding us within its essence and reminding us of our place in the infinite whole. Navigating existence with this understanding, we glimpse fragments of the vast expanse, humbled and inspired by the connection to the boundless universe.

VI. The Order and Chaos of the Universe

The fundamental order of the universe is both a scientific and philosophical mystery. Questions arise: What keeps the universe from imploding or flying apart? Who, or what, maintains the cosmos? Is there a custodian overseeing each molecule, each atom, or do the laws of physics themselves sustain everything? Science suggests that gravity, electromagnetism, and nuclear forces work together to maintain order, preventing collapse or chaos. Yet, the deeper source and nature of this order remain elusive.

At the microscopic level, quantum mechanics provides a framework for understanding particles, but the origin of these laws is still unknown. The energy that fuels the universe is another profound mystery—while the Big Bang theory posits a beginning, the ultimate source of this energy is speculative. Considering whether the universe is order or chaos, one might reflect that what appears chaotic on one level may be higher-order organization on another. Thus, the full nature of the universe remains an awe-inspiring mystery, inviting continuous exploration and wonder.

VII. The Egg as Concept and Genesis

The egg is a concept that predates its nestlings—beings of lesser nature, destined to grow in awareness and eventually question their existence. Yet, the intellect that spawned the concept of the egg must have been more advanced than the egg itself, suggesting a hierarchy of creation. The egg represents the beginning, pure and unaffected, emerging from the shadows of non-existence to be given form and being.

In the grand narrative of existence, the egg symbolizes the inception of all things, a primordial concept preceding its own creations. The beings that emerge from the cosmic egg embark on a journey of growth and self-awareness, evolving until they contemplate their origins and purpose. The nature of the egg is shrouded in mystery, its originator a transcendent force beyond the egg itself. The egg exists in a state of primordial purity, the initial spark from which the tapestry of existence unfolds.

By viewing the egg in this light, we gain insight into beginnings and the unfolding of reality. Every complex structure traces its origin to a simple, unblemished state of potential. The journey from the egg to awareness testifies to the transformative power of existence—where simplicity gives rise to complexity and the unformed becomes the foundation of all being.

VIII. Mystical Concepts and Meta-Reality

Is there something outside this reality that creates the mystical concepts we perceive? Are these ideas passing fancies or part of a stream of motion that neither began here nor will end here? Contemplating our reality and mystical concepts leads us to question the existence of something beyond our universe—a meta-reality from which profound ideas and experiences might originate.

The notion that our reality could be part of a larger system suggests that mystical concepts may transcend our ordinary understanding, possibly originating from a realm beyond our own. This higher plane might be the birthplace of archetypes, symbols, and ideas that permeate consciousness, offering glimpses of deeper truths. Our stream of motion—life, time, and unfolding events—may be but one phase in an endless continuum, a fleeting moment in an infinite flow.

Thus, our lives become part of a larger, ongoing story, with mystical concepts serving as echoes of a deeper reality. These experiences connect us to infinity, suggesting that our existence is a thread in the universe's vast tapestry. The interplay between known and unknown, finite and infinite, encourages us to appreciate the profound interconnectedness of all things and to transcend the limitations of our current understanding.

IX. The Alien Nature of the Egg

Is the egg an alien concept? If so, what force or consciousness formed it? The question of the egg's origin is a final mystery, inviting speculation about the nature of the intelligence or power that conceived it. The egg may be more than a symbol; it could be the product of an otherworldly force or consciousness, its existence pointing to realities and powers beyond ordinary comprehension.

Some mystics believe that life on this planet is the direct result of extraterrestrials that seeded this world billions of years ago. Much like those who profess that the race of luminous beings known as the Elohim brought forth life on this planet. Many books have been written, ranging from theology and mythology to science fiction and facts. The truth is that one must believe what one believes is the truth.

This notion, often referred to as panspermia or directed panspermia, suggests that life on Earth was seeded by advanced extraterrestrial beings. Proponents of this idea point to various ancient texts, mythologies, and even modern scientific theories to support their claims. The Elohim, described in some religious traditions as powerful, luminous beings, are one such example of entities believed to have played a role in the creation and development of life on Earth.

The theory spans a wide array of disciplines and interpretations:

- Many ancient cultures have stories of gods or celestial beings who came to Earth and influenced the development of human civilization. These narratives often describe the Elohim or similar beings as teachers, creators, or overseers of humanity.
- Writers and thinkers have long explored the idea of extraterrestrial life and its impact on human history. Works of science fiction often imagine advanced alien civilizations with the capability to seed life on other planets, providing a creative and speculative lens through which to view these possibilities.
- While mainstream science generally attributes the origins of life to natural processes on Earth, the panspermia hypothesis remains a topic of interest. Some scientists propose that life could have been transported to Earth via comets, asteroids, or even deliberate actions by advanced extraterrestrial civilizations.

Ultimately, the question of our origins is deeply personal and philosophical. The vast array of theories and beliefs reflects the complexity of this question and the diverse ways in which humans seek to understand their place in the universe.

The statement "one must believe what one believes is the truth" underscores the subjective nature of belief and truth. Each individual's understanding of reality is shaped by their experiences, knowledge, and worldview. While empirical evidence and scientific inquiry strive to provide objective truths, personal beliefs and interpretations play a significant role in how we perceive and make sense of the world.

In the end, the search for truth is a journey that each

person undertakes in their own way. Whether through science, spirituality, mythology, or a combination of these, the quest to understand our origins continues to inspire and challenge us, driving us to explore the mysteries of existence and our place within the cosmos.

X. Logic, Metaphysics, and the Nature of Reality

And since logic cannot exist without structure or the calculated progression of ideas, any formal discussion of reality that ventures into the area of metaphysics most certainly will be skewed by subjective belief. Yet, the beginning point of any truly scientific discussion must begin with the premise that reality exists. Thus, we endeavor to understand this phenomenon of how anything exists. In the realm of metaphysics, where the nature of reality and existence itself is questioned, the interplay between logic and subjective belief becomes particularly pronounced.

Logic serves as the backbone of rational inquiry. It establishes a framework within which ideas can be tested, validated, or refuted based on consistent principles. In scientific discussions, logic helps to build models of understanding that can be observed and experimented upon. However, metaphysical inquiries often extend beyond the reach of empirical validation, venturing into territories where logic must coexist with speculation and belief.

When discussing metaphysical questions—such as the nature of existence, the origin of the universe, or the existence of a higher consciousness—subjective beliefs inevitably color the conversation. Personal experiences, cultural background, religious beliefs, and philosophical outlooks shape how individuals perceive and interpret these profound questions. This subjectivity can lead to diverse and often conflicting viewpoints, each offering a unique perspective on the nature of reality.

Despite the subjective influences, any scientific exploration of reality must start with a fundamental premise: that reality, in some form, exists. This assumption is the cornerstone of all scientific inquiry. By accepting that there is an objective reality to be understood, scientists can employ logical reasoning and empirical methods to explore and explain the phenomena they observe.

Understanding how anything exists is one of the most profound and challenging questions in both science and philosophy. This endeavor involves exploring cosmology, quantum physics, philosophy of mind, and metaphysics. Cosmology is the study of the origin, structure, and dynamics of the universe. The Big Bang theory, cosmic inflation, and the nature of dark matter and dark energy are key areas of exploration. Quantum Physics involves investigating the fundamental particles and forces that constitute reality at the smallest scales. Quantum mechanics challenges classical notions of reality, introducing concepts like superposition and entanglement. Philosophy of Mind relates to examining the nature of consciousness and its relationship to the physical world. Questions about the mind-body problem, the nature of self-awareness, and the possibility of an afterlife are central to this field. Metaphysical inquiry often intersects with theology, exploring concepts like the existence of a higher power or the nature of the soul.

The pursuit of understanding existence is a multifaceted endeavor that blends logic, empirical evidence, and subjective belief. While logic and structure provide the tools for rational exploration, subjective beliefs offer diverse perspectives that enrich the discourse. By starting with the premise that reality exists, scientists and philosophers alike can embark on a journey to unravel the mysteries of existence, continually expanding our understanding of the universe and our place within it.

XI. The Egg and the Atomic Universe

Let us then return to the concept of the egg. If it is indeed valid, then it will grow. We have discussed how reality exists in multiple layers, expanding both outward and inward simultaneously. We have discussed how each layer reflects the next, and so on. This can be illustrated more clearly in the example of the atomic universe.

The egg, as a symbol of potential and genesis, represents the starting point from which all complexity arises. If this concept is valid, it follows a natural progression of growth and expansion. Just as a biological egg contains the potential for life and development, the metaphysical egg symbolizes the origin of existence, unfolding into increasingly complex forms.

Reality can be viewed as consisting of multiple layers, each reflecting and influencing the others. These layers can be thought of as a nested hierarchy, where each level encompasses and interacts with the levels both above and below it. This nested structure is a fundamental characteristic of the universe, observable in various domains from the microscopic to the macroscopic.

One of the most illustrative examples of this layered reality is the atomic universe. Atoms, the building blocks of matter, provide a microcosmic view of the layered structure of reality. At the most fundamental level, atoms are composed of subatomic particles—protons, neutrons, and electrons. These particles themselves are made up of even smaller constituents, such as quarks and leptons, governed by the fundamental forces of nature. These subatomic particles come together to form atoms, which are the basic units of chemical elements. Each atom consists of a nucleus (containing protons and neutrons) surrounded by a cloud of electrons.

Atoms bond together to form molecules, which are the chemical compounds that make up the substances we encounter in everyday life. The interactions between atoms within molecules are governed by chemical bonds and physical laws. On a larger scale, molecules assemble to form the structures within living cells. Cells, in turn, organize into tissues, organs, and entire organisms, each level adding complexity and functionality. Organisms interact within ecosystems, and ecosystems form part of the biosphere, which integrates with the planet and the broader cosmos.

Each layer of reality reflects the structure and dynamics of the layers above and below it. For instance, the behavior of molecules is influenced by the properties of the atoms that constitute them, while the functionality of cells depends on the molecular interactions within them. This recursive structure is mirrored across different scales, from the quantum realm to the cosmic expanse.

Just as the egg grows and evolves, so too does each layer of reality. The universe itself can be seen as an ongoing process of growth and development, where complexity emerges from simplicity through a series of transformative steps. This evolutionary process is driven by underlying principles and forces that govern the interactions and relationships between different layers. The concept of the egg, as a symbol of origin and potential, encapsulates the idea of growth and expansion across multiple layers of reality. By examining the atomic universe, we gain insight into the nested structure of existence, where each layer reflects and influences the others. This layered reality, expanding both outward and inward, illustrates the dynamic and interconnected nature of the cosmos, continually evolving and unfolding from its primordial beginnings.

XII. The Macrocosm and Microcosm

Imagine the planets in our solar system moving about the sun in their respective orbits. Jupiter and Saturn have multiple planetoids or moons orbiting them as well. Our sun is one of many billions of stars in the galaxy; billions of stars that hold planets and moons in constant orbit. And of course, there are billions of galaxies in the universe. As you see, this concept proceeds along the path toward the macrocosm of the egg.

The egg, as a symbol of origin and potential, expands from the microcosmic to the macrocosmic, illustrating the interconnectedness and vastness of the universe. By following this conceptual path, we can appreciate how the intricate dance of celestial bodies reflects the deeper principles of cosmic order and growth.

Planets, including Earth, orbit the sun in elliptical trajectories, governed by gravitational forces. Each planet follows a unique path, maintaining a delicate balance of motion. Larger planets like Jupiter and Saturn host numerous moons and planetoids, each orbiting their parent planet. These moons themselves may have complex systems, including subsurface oceans and geological activity.

Our sun is just one of approximately 100 billion stars in the Milky Way. Each star may possess its own system of planets and moons, creating a vast network of stellar and planetary interactions. The Milky Way itself is a barred spiral galaxy, with stars, gas, and dust organized into spiral arms. At the center lies a supermassive black hole, around which the galaxy rotates.

The universe contains an estimated 2 trillion galaxies, each with its own unique structure and composition. These galaxies range from spirals and ellipticals to irregular shapes, each a cosmic egg of potential. Galaxies group together into clusters, and these clusters form even larger structures known as superclusters. These vast arrangements create a cosmic web, with filaments of dark matter and intergalactic gas connecting them.

This progression from the solar system to the universe illustrates the concept of the macrocosm of the egg. Each layer of reality, from planets orbiting stars to galaxies forming clusters, reflects a nested structure of increasing complexity.

This mirrors the growth and expansion inherent in the egg concept, where potential unfolds into diverse and intricate forms.

The egg, therefore, serves as a powerful metaphor for the universe's unfolding complexity. It encapsulates the idea that from a singular point of origin—whether a cosmic egg or a primordial singularity—the universe has expanded and diversified into a vast and interconnected macrocosm. By imagining the celestial bodies and their orbits, we trace a path from the microcosm to the macrocosm, revealing the layered structure of reality. The egg, as a symbol of potential and growth, guides our understanding of this cosmic order, illustrating how the universe unfolds in ever-expanding layers of complexity. This journey from the smallest scales to the grandest expanses highlights the profound interconnectedness of all things, encouraging us to explore and appreciate the intricate beauty of the cosmos.

XIII. The Microcosm

Now imagine a simple atom of oxygen, which contains eight small electrons in constant orbit around its immense nucleus of protons and neutrons. Here, as you can tell, we are proceeding down the path toward the microcosm of the egg.

Just as the egg symbolizes growth and potential on the macrocosmic scale, it also serves as a metaphor for the intricate and nested structures found within the microcosm. By exploring the atomic and subatomic realms, we can better understand the foundational layers of reality.

At the atomic scale, consider the oxygen atom, which provides a clear example of the microcosm: Oxygen has eight electrons that orbit its nucleus in defined energy levels or shells. These electrons are in constant motion, creating a dynamic and complex electron cloud. The nucleus of an oxygen atom contains eight protons and eight neutrons. Protons are positively charged particles, while neutrons have no charge. Together, they form the dense core of the atom.

Diving deeper into the microcosm, we encounter the realm of subatomic particles. Both protons and neutrons are composed of even smaller particles called quarks, held together by the strong nuclear force. Protons contain two "up" quarks and one "down" quark, while neutrons contain two "down" quarks and one "up" quark. Quarks are fundamental particles that interact via the exchange of gluons, the carriers of the strong force. This interaction binds quarks together to form protons and neutrons.

At this level, the principles of quantum mechanics come into play: Electrons and other subatomic particles exhibit both wave-like and particle-like properties. This duality is a key feature of the quantum world. Heisenberg's uncertainty principle states that we cannot simultaneously know the exact position and momentum of a particle. This introduces a fundamental limit to our precision in measuring the properties of subatomic particles.

The concept of the egg in the microcosm illustrates how complexity emerges from simplicity at the smallest scales. Just as the macrocosm consists of nested layers of celestial bodies, the microcosm contains nested structures of particles within atoms. Each layer builds upon the previous one, creating a hierarchy of matter.

The egg metaphor applies to the microcosm, symbolizing the potential for growth and complexity within the atomic and subatomic realms. From fundamental particles to complex atoms, the microcosm unfolds in a manner analogous to the macrocosmic expansion. The microcosm mirrors the macrocosm in its layered complexity and dynamic interactions. The behavior of electrons orbiting a nucleus reflects the motion of planets around a star, highlighting the fractal-like nature of reality. Each layer, whether cosmic or atomic, follows principles that govern its structure and interactions.

By imagining the structure of an oxygen atom, we delve into the microcosm of the egg, revealing the intricate and nested layers of reality at the smallest scales. This exploration underscores the profound interconnectedness and complexity of the universe, from the vast expanses of galaxies to the minute realms of subatomic particles. The egg, as a symbol of origin and potential, encapsulates the growth and unfolding of complexity across all scales of existence.

XIV. Reflection and Symmetry

The solar system and the atom have common traits and can be envisioned as reflections of one another. The universe we live in can, therefore, be perceived as either very great or exceedingly small, depending upon your beliefs.

Both the solar system and the atom exhibit similar structural characteristics, demonstrating the fractal nature of the universe. By examining these common traits, we can gain a deeper appreciation of the interconnectedness and symmetry inherent in the cosmos.

At the center of our solar system lies the sun, around which planets and other celestial bodies orbit. The gravitational force of the sun governs the motion of these orbiting bodies. At the heart of an atom is the nucleus, composed of protons and neutrons. Electrons orbit the nucleus in defined energy levels, influenced by the electromagnetic force.

Planets may have their own moons, creating a nested system of orbits within the larger solar system. These moons orbit their parent planets, which in turn orbit the sun. Electrons occupy different energy levels or shells around the nucleus of the atom. These shells create a hierarchical structure, with electrons filling lower energy levels before higher ones. The planets and moons are in constant motion, following elliptical orbits around the sun and their respective planets. Electrons are in perpetual motion, creating an electron cloud that represents probabilistic locations of electrons rather than fixed orbits.

Depending on one's perspective, the universe can be viewed as either immensely vast or incredibly minute. This duality highlights the relativity of scale and the profound nature of existence.

From the macroscopic perspective, the universe is an expansive, awe-inspiring realm filled with countless galaxies, stars, and planetary systems. This view emphasizes the vastness and grandeur of the cosmos. The solar system, as a part of this immense universe, represents just one of many intricate systems, each contributing to the overall complexity and beauty of the cosmos. From the microscopic perspective, the universe is an intricate web of particles and forces, each playing a crucial role in the fabric of reality. This view underscores the detailed and delicate nature of the atomic and subatomic realms.

The atom, as a fundamental unit of matter, mirrors the structure and dynamics of larger systems, suggesting that the principles governing the cosmos are consistent across different scales. The similarities between the solar system and the atom illustrate the concept of reflection and symmetry in nature. This fractal-like pattern suggests that the universe is built upon recurring principles, manifesting in both the largest and smallest scales. By understanding these patterns, we can gain insights into the fundamental nature of reality.

Ultimately, whether the universe is perceived as very great or exceedingly small depends on individual beliefs and perspectives. This duality reflects the infinite complexity and wonder of the cosmos, inviting us to explore and appreciate the interconnected layers of existence. The solar system and the atom, with their common traits and reflective structures, exemplify the fractal nature of the universe. Depending on one's viewpoint, the universe can be seen as either a vast expanse or a delicate microcosm. This dual perspective highlights the profound interconnectedness and symmetry that define our reality, encouraging us to marvel at the intricate beauty of the cosmos.

XV. Layers and Multi-Dimensional Reality

Layers ... multi-dimensional levels of reality expanding into infinity, like eggs, growing ... each with its purpose and design. If the egg truly is a valid concept, we shall conceive one that will grow into realizing the circumstances and basis for existence. When fully matured, our concept will unfold an explanation that logic has failed to uncover.

The concept of layers and multi-dimensional reality suggests that existence is composed of an infinite number of levels, each expanding and evolving like an egg. Each layer or dimension has its own unique purpose and design, contributing to the overall complexity and harmony of the universe.

The egg, as a metaphor, represents the potential for growth and the unfolding of intricate structures from a simple origin. This idea can be applied to the multi-dimensional layers of reality, where each "egg" grows and evolves, revealing new levels of complexity and purpose.

The notion of layers expanding into infinity implies that the universe is boundless and ever evolving. Each layer of reality, whether it be a physical, metaphysical, or spiritual dimension, contributes to the infinite expansion of existence. This continuous growth mirrors the biological process of an egg developing into a fully formed organism.

Each layer of reality has its own unique purpose and design, much like the components of an egg that develop into different parts of a living being. These layers work together in harmony, creating a cohesive and interconnected system. Understanding the purpose and design of each layer can provide insight into the overall structure and function of the universe.

If we accept the egg as a valid concept, we can conceive a growing idea that helps us understand the circumstances and basis for existence. This concept, like an egg, will develop and expand, revealing deeper truths and explanations that traditional logic may not fully encompass.

As our concept matures, it will unfold an explanation for existence that transcends conventional logic. This unfolding process is akin to the development of an egg, where each stage of growth reveals new aspects and dimensions. Ultimately, this matured concept will provide a comprehensive understanding of existence, addressing questions and mysteries that logic alone cannot solve.

The metaphor of the egg, with its layers and multi-dimensional levels of reality, offers a profound framework for understanding the universe. By embracing this concept, we can conceive a growing idea that will expand our understanding of existence. As this concept matures, it will reveal explanations and insights that transcend traditional logic, providing a deeper and more holistic comprehension of the circumstances and basis for our reality.

XVI. Conceptual Seeds

Here, we will plant a **conceptual seed**, for we believe concepts exist as living things ... spawning into dynamic forces rather than mere inanimate objects. Because they are alive, they will grow and mature. Like all living things, they must be embraced and nurtured. Thus, our conceptual seed will fall upon **fertile soil** or **rocky ground**. The fertile soil will accept and nurture the seed, and it will grow to maturity and bear fruit. The rocky ground, however, will reject the seed and lay wasting and withering into decay.

The idea of planting a conceptual seed emphasizes that concepts are not static entities but dynamic forces that have the potential to grow and evolve. By treating concepts as living things, we acknowledge their capacity for development and transformation.

Concepts are not merely abstract ideas; they are dynamic forces that can influence and shape our understanding and actions. Just like living organisms, concepts have the potential to adapt, evolve, and expand over time.

When nurtured and embraced, concepts can grow to maturity, developing into fully realized ideas that can bear fruit in the form of knowledge, innovation, and understanding. This process of growth parallels the biological development of living beings, highlighting the organic nature of conceptual evolution. For a conceptual seed to thrive, it must be nurtured and cared for. This involves providing the necessary conditions for growth.

Fertile soil represents an open and receptive mind, willing to engage with new ideas and nurture their development. In such an environment, the conceptual seed can take root, grow, and eventually bear fruit, contributing to a deeper understanding and new perspectives.

Rocky ground symbolizes a closed or resistant mindset, where new ideas are rejected or ignored. In such conditions, the conceptual seed cannot thrive, and it will ultimately wither and decay, failing to reach its potential.

To ensure that our conceptual seeds fall upon fertile soil, we must cultivate openness and support. Fostering an environment that encourages curiosity, critical thinking, and a willingness to explore new ideas is critical for intellectual growth. Engaging in discussions and interactions that challenge existing beliefs promotes the growth of innovative concepts. Providing resources, guidance, and encouragement help concepts develop and mature.

Creating spaces where ideas can be tested, refined, and expanded allows them to reach their full potential. By planting and nurturing conceptual seeds, we can cultivate a rich and dynamic intellectual landscape. Concepts, like living things, have the potential to grow, evolve, and bear fruit when provided with the right conditions. Embracing and nurturing these seeds ensures that they fall upon fertile soil, leading to the growth of knowledge, innovation, and understanding. Conversely, if they fall upon rocky ground, they will wither and decay, leaving untapped potential unrealized. Through mindful cultivation, we can ensure that our conceptual seeds thrive and contribute to the flourishing of ideas and insights.

XVII. Folklore, Superstition, and the Cosmos

You will plant the conceptual seed when the new moon rises, and you will reap the true harvest when the moon is full. Folklore and superstition carried many misgivings about the nature of the universe and the phenomenon of reality.

The symbolism of planting a conceptual seed at the new moon and reaping the harvest at the full moon suggests a cycle of growth and maturation. This metaphor highlights the phases of development that ideas undergo, from inception to fruition. The new moon represents a time of new beginnings and potential. It is a moment of darkness, where the seed is planted, symbolizing the start of an idea or concept. In this phase, the idea is fragile and requires care and nurturing to ensure it takes root and begins to grow.

Planting the seed involves setting intentions and envisioning the potential of the concept. It is a time for thoughtful planning and preparation. This stage is crucial for establishing the foundation upon which the concept will develop. The full moon signifies the culmination of the growth cycle, where the concept has reached its full maturity. It is a time of illumination and realization, where the results of nurturing and effort become visible.

The harvest represents the tangible outcomes of the conceptual seed, bearing fruit in the form of insights, knowledge, and achievements. Reaping the harvest is also a time for reflection on the journey of growth and the lessons learned along the way. It is a moment to express gratitude for the process and the resulting fruits, acknowledging the effort and support that contributed to the concept's maturation.

Throughout history, folklore and superstition have shaped human understanding of the universe and the nature of reality. These beliefs often carried misgivings and misconceptions, reflecting the limitations of knowledge at the time. Folklore encompasses traditional stories, myths, and cultural narratives that explain natural phenomena and human experiences. While rich in symbolism and cultural significance, folklore often includes fantastical elements that deviate from scientific explanations. These stories served to provide meaning and understanding in a pre-scientific world, offering comfort and a sense of order.

Superstition involves beliefs and practices based on fear of the unknown and the desire to influence outcomes through ritualistic behaviors. Misgivings about the nature of the universe and reality often stemmed from a lack of empirical knowledge, leading to the creation of superstitions to explain and control the world around us.

Modern science and philosophy have advanced our understanding of the universe, moving beyond the misgivings of folklore and superstition. However, the symbolic and metaphorical wisdom embedded in these traditions can still offer valuable insights:

- The metaphors and symbols in folklore and superstition can provide a deeper, more intuitive understanding of human experiences and the natural world. These narratives can enrich our appreciation of the universe by connecting us with cultural heritage and collective wisdom.

- By integrating scientific knowledge with the symbolic insights of folklore, we can develop a more holistic understanding of reality. This approach allows us to honor the past while embracing the advancements of the present, creating a balanced perspective on the nature of the universe.

Planting a conceptual seed at the new moon and reaping the harvest at the full moon represents the journey of growth and realization. While folklore and superstition have historically shaped our understanding of reality, modern knowledge allows us to move beyond these misgivings. By integrating the symbolic wisdom of the past with contemporary insights, we can cultivate a deeper and more nuanced appreciation of the universe and the phenomenon of reality.

XVIII. Growth and Awakening

Early in mankind's history, it was believed that the world was flat, that it was the center of the universe, and that the stars, the sun, and the moon were gods that traveled from horizon to horizon above this mortal stage. Scholars, philosophers, clergy, and potentates honestly believed this was reality. Anyone who dared to believe otherwise was branded a fool or a heretic.

In the early stages of human civilization, our understanding of the universe was limited by the tools and knowledge available at the time. These early beliefs, though now disproven, were foundational to the cultural and intellectual development of humanity. The flat Earth model was a common belief across many ancient cultures. It was a natural conclusion drawn from the everyday experience of seeing a seemingly flat horizon. This belief was reinforced by mythology and religious teachings that depicted the Earth as a flat plane, often supported by a cosmic structure or deities. Many ancient civilizations, including the Mesopotamians, Egyptians, and early Greeks, subscribed to the flat Earth model. These beliefs were deeply ingrained in their cosmologies and worldviews.

Geocentrism, the belief that the Earth was the center of the universe, was widely accepted in ancient and medieval times. This model posited that all celestial bodies orbited the Earth. The geocentric model was supported by the teachings of influential philosophers like Aristotle and Ptolemy, who argued that the Earth's central position was evident from the observable motion of the stars and planets. Many religions endorsed the geocentric view, seeing it as aligning with theological doctrines that placed humanity at the center of creation. This belief was not just scientific but also deeply intertwined with religious and existential perspectives.

In many ancient cultures, celestial bodies were personified and worshiped as gods or divine beings. The predictable movements of the sun, moon, and stars were seen as the actions of these deities. These beliefs were central to the mythologies and religious practices of civilizations like the Greeks, Romans, Egyptians, and Mesopotamians. Myths and legends described the journeys of these celestial gods across the sky, influencing the natural world and human affairs. These narratives provided explanations for the cycles of day and night, seasons, and other astronomical phenomena.

The deeply held beliefs about the cosmos were fiercely defended by the scholarly, religious, and political authorities of the time. Challenging these views was seen as an affront to established knowledge and divine order. Those who proposed alternative models, such as a spherical Earth or a heliocentric (sun-centered) universe, faced severe opposition. Individuals who dared to question the prevailing cosmological models were often ridiculed, persecuted, or branded as heretics. This included notable figures like Galileo Galilei, who supported the heliocentric model proposed by Copernicus. The clash between new ideas and established beliefs highlights the tension between innovation and tradition in the history of human thought.

The early beliefs about a flat Earth, a geocentric universe, and celestial bodies as gods reflect the limited knowledge and cultural contexts of ancient civilizations. These models were grounded in observation, mythology, and religious doctrine. However, as human understanding evolved, so too did our models of the cosmos. The transition from these ancient beliefs to modern scientific understanding was marked by significant resistance and controversy, underscoring the challenges of intellectual progress. Today, we recognize the importance of questioning and revising our understanding in light of new evidence, a principle that continues to drive scientific and philosophical advancements.

XIX. God-Consciousness and Oneness

"We are eggs of a lesser nature, and we are growing. Each life is experienced in a series of stages... birth to death and death to beyond. We reach forward and into the past for enlightenment. The ancient ones... our ancestors, consciously aware of existence long before our concepts were in play, possess wisdom perceptible beyond anything we can imagine. Thus, we seek... asking for protection and guidance to the upper levels of conscious awareness, where we can awaken from this psychic slumber."

The metaphor of being "eggs of a lesser nature" emphasizes our potential for growth and transformation. Just as an egg contains the potential for a fully formed being, we too possess the capacity for significant development. This growth is not just physical but also spiritual and intellectual.

Life is experienced in a series of stages, from birth to death and beyond. Each stage represents a phase in our development, offering opportunities for learning and enlightenment. The transition from one stage to another can be seen as a form of rebirth, where death in one phase leads to new experiences and growth in the next.

We reach forward into the future and backward into the past in our quest for enlightenment. This dual movement underscores our desire to learn from both future possibilities and ancient wisdom. The knowledge and experiences of our ancestors provide a foundation upon which we build our understanding of existence.

Our ancestors, the "ancient ones," were consciously aware of existence long before our modern concepts emerged. Their wisdom, accumulated over generations, holds insights that can guide us in our journey. This wisdom is often perceived as being beyond our current understanding, suggesting that the past holds profound truths that we are yet to fully comprehend.

In our journey towards higher consciousness, we seek protection and guidance from the ancient ones. This reflects a recognition of our vulnerability and the need for support in navigating the complexities of existence. Asking for guidance implies humility and openness to learning from those who have come before us.

The goal is to ascend to the upper levels of conscious awareness, where we can awaken from our psychic slumber. This awakening represents a heightened state of understanding and enlightenment. Achieving this state requires effort, introspection, and a willingness to transcend our current limitations.

Psychic slumber refers to a state of unawareness or limited consciousness. It implies that while we may be physically awake, we are not fully conscious of the deeper truths of existence. Awakening from this slumber involves expanding our awareness and gaining a deeper understanding of ourselves and the universe.

The journey to awakening is a process of self-discovery and growth. It requires us to confront our limitations, seek out wisdom, and remain open to new insights. This journey is both individual and collective, as we learn from our own experiences and the shared knowledge of humanity.

The metaphor of being "eggs of a lesser nature" captures the essence of human potential and the stages of growth we experience. By reaching forward into the future and backward into the past, we seek enlightenment and wisdom from our ancestors. In this quest, we ask for protection and guidance to ascend to higher levels of conscious awareness. Awakening from our psychic slumber is the ultimate goal, representing a state of heightened understanding and enlightenment. This journey requires humility, openness, and a deep connection to the wisdom of those who have come before us.

"When we awaken, we shall know God-consciousness. And as eggs of a slightly greater nature, we shall grow into the oneness of His Being."

Awakening to God-consciousness represents reaching a state of ultimate awareness and unity with the divine. It is the realization of a higher truth and the understanding of our connection to a greater spiritual reality. This state transcends ordinary consciousness, allowing us to perceive and experience the divine presence in all aspects of existence. The process of awakening is a transformative journey. It involves shedding layers of ignorance and illusion, and embracing a deeper, more profound understanding of our true nature. This journey is marked by spiritual growth, self-discovery, and the cultivation of inner wisdom.

As we awaken, we evolve from being "eggs of a lesser nature" to "eggs of a slightly greater nature." This signifies our ongoing growth and development on the spiritual path. This progression reflects our increasing capacity to understand and embody divine qualities. Each stage of our spiritual journey brings us closer to the realization of our divine potential. Just as an egg develops into a fully formed being, our spiritual growth leads us towards a more complete and unified state of existence.

Growing into the oneness of His Being signifies the ultimate goal of our spiritual journey: unity with the divine. This oneness is characterized by a profound sense of connection and harmony with God and the universe. It is the realization that we are not separate from the divine, but rather an integral part of it. Achieving oneness with His Being involves the full realization of our divine nature. It is the understanding that our true self is inherently connected to God, and that we are expressions of divine consciousness. This realization brings about a deep sense of peace, love, and purpose.

The journey towards awakening and God-consciousness is a transformative process of spiritual growth and self-discovery. As we evolve from "eggs of a lesser nature" to "eggs of a slightly greater nature," we move closer to realizing our unity with the divine. This path leads us to the ultimate goal of oneness with His Being, where we fully embody and understand our divine nature. Through this journey, we transcend ordinary consciousness and awaken to a higher truth, experiencing the profound connection and harmony of God-consciousness.

CHAPTER TWO

The Infinite Mirror

The Universe is So Large that Language Breaks Against It.

I. The Universe as One Body

The mind tries to visualize scale—the distances between galaxies, the age of stars, the indifferent geometry of space—and then it reaches for the word infinite. Often the word is not a measured conclusion. It is a confession: the person has reached the limit of imagination.
But a limit can be used. When a mind encounters a boundary, it can retreat into cynicism—"it is unknowable"—or it can cultivate reverence—"it is larger than the current model." Reverence is not blind belief. It is the discipline of not confusing the present model with the whole.

The esoteric tradition insists that the universe is one entity: a single living system with many expressions. That insistence is not merely poetic. It is a method of orientation. If everything is connected, then a person's attention becomes morally and spiritually consequential. If everything is connected, then the boundary between "inner" and "outer" becomes less absolute.

Modern science, in its own vocabulary, also describes interconnection—fields, forces, causal webs, ecological systems. It does not declare the universe "one mind," yet it repeatedly shows that nothing stands alone. In that sense, esoteric unity and scientific interdependence are not enemies; they are different styles of speaking about relationship.

The book begins from this orientation: reality is not a pile of separate things, but a system. The human being is not a detached spectator, but a subsystem that can become self-aware.

When awareness turns inward, the person does not leave the universe. The person approaches the universe from another side.

Science Note: Cosmology provides models of large-scale structure and interdependence without implying consciousness as a fundamental property. Some philosophical positions (e.g., panpsychism) explore such implications, but they remain debated. The book uses "one body" primarily as an orienting metaphor, not as a testable claim.

II. The Human Body as a Cosmos

To call the human body a cosmos is not flattery. It is anatomical accuracy expressed in mythic language. The body is a network of specialized systems coordinated through flow—blood, lymph, hormones, neural signaling. It is a city of cellular citizens governed not by a single ruler but by distributed regulation.

The ancient phrase "microcosm" points to a practical truth: systems within systems. The heart is a pump, yes, but also an instrument that responds to emotion, posture, exertion, and breath. The lungs are bellows, yes, but also a doorway into autonomic regulation. The gut is digestion, yes, but also a vast neural and microbial ecosystem that shapes mood and cognition.

A person lives in this cosmos most of the time without noticing it. The body becomes background. The mind uses it as a vehicle and calls it "me," then forgets to listen.

Psychic awakening begins, paradoxically, with embodiment. Without the capacity to settle into breath and sensation, inner exploration becomes fantasy rather than perception. The esoteric traveler who cannot regulate fear, excitement, and attention will confuse noise for revelation.

The body is not a cage. It is a tuning fork. The more precisely it is tuned—sleep, rhythm, breath, honest self-observation—the more clearly the inner worlds can be entered without distortion.

Science Note: Neuroscience treats consciousness as emerging from coordinated activity across distributed networks rather than a single "seat." Interoception (perception of internal bodily signals) is increasingly recognized as central to emotion and selfhood. Practices that increase bodily awareness can affect autonomic regulation and emotional stability.

III. Tools, Generations, and the Accumulating Mind

Each generation inherits tools and stories. It inherits language, calendars, myths, measurement devices, and the half-solved problems of its ancestors. It also inherits blind spots. The human species does not progress in a straight line; it loops, forgets, rediscovers, and renames.

The modern world privileges external tools—technology, computation, networks—because external tools produce visible change. The esoteric world privileges internal tools—attention, intention, symbol—because internal tools produce invisible change that reshapes perception itself.

Both toolsets matter. A telescope extends the eye, but attention determines whether a person ever looks through it. A brain scanner can map neural correlates, but it cannot substitute for the lived experience of waking up inside a dream. A smartphone can store information, but it cannot store wisdom.

The deepest tool is training. Training is what converts potential into function. Psychic slumber persists not because people lack potential but because they rarely train the faculties that perceive the inner planes: memory for dreams, metacognition, emotional regulation, symbolic fluency, ethical grounding, and sustained intention.

This book treats training as sacred and practical. It assumes no single belief system is required. It insists only on discipline: the willingness to test, record, refine, and return.

Science Note: Skill acquisition research emphasizes deliberate practice, feedback, and consistency. Lucid dreaming training, likewise, tends to improve with systematic journaling, intention-setting, and iterative refinement. The book's "internal tools" correspond to measurable cognitive processes such as attention, working memory, and metacognition.

IV. The Egg Grows While Humanity Sleeps

In the older teaching, there is an image: an egg growing while humanity sleeps. The egg represents latent capacity—an inner body, an unhatched self, an expanded mode of awareness that develops quietly even when ignored. Whether one imagines this as a literal astral body or as a metaphor for cognitive and spiritual maturation, the message is the same: potential does not guarantee awakening.

Many people sense the egg indirectly. They feel a pressure of meaning that cannot be satisfied by routine. They feel an ache that consumption does not cure. They feel a haunting sense that something vital is being missed.

This sense is often medicated by distraction. In the modern world distraction is not merely entertainment; it is a sophisticated economy. The attention of a person is valuable, and industries compete to capture it. The mind learns to scroll, react, compare, and crave. Over time the nervous system begins to prefer stimulation over stillness.

Stillness, however, is where the egg becomes visible. In stillness the mind confronts itself. It must face desire without gratification, fear without immediate solution, boredom without escape. Most people avoid this confrontation for years. Then they wonder why life feels thin.

The esoteric path does not condemn ordinary pleasures. It simply refuses to allow them to become a substitute for awakening. A person can enjoy the world and still remain asleep. The goal is not austerity. The goal is lucidity—waking life lived with the same clarity sought in the dream.

Science Note: Research on attention and digital distraction suggests that constant novelty and reward cues can shape attentional habits and reduce tolerance for boredom. While claims about "spiritual thinness" are not scientific categories, the described pattern resembles reinforcement learning dynamics and attentional conditioning.

V. The Psychic Tomb: Diversions of the Waking World

Psychic slumber has gatekeepers, and they rarely appear as villains. They appear as reasonable priorities: success, recognition, security, status, romance, the pursuit of pleasure, the avoidance of pain. These are not inherently corrupt. They become tombstones when they are treated as ultimate.

A tomb is not always built from stone. It can be built from routines that never question themselves. It can be built from social scripts that substitute performance for authenticity. It can be built from internal narratives—"this is who I am"—that harden into identity and resist growth.

The older teaching warns of fame, glory, riches, power, sex, and love as diversions. The modern mind may bristle at this list, especially at love. Yet the teaching is not anti-love. It is anti-capture. Even love can become a diversion when it is used to avoid the deeper work of facing the self. Love can become a bargaining chip for validation. Sex can become a tranquilizer. Power can become an anesthetic. Money can become a substitute for meaning.

Awakening requires a different relationship to these forces: not rejection, but sovereignty. The person must become capable of choosing rather than being dragged.

This is the moral ground of lucid practice. The same skills used to control dreams—expectation, intention, narrative manipulation—can also be used to manipulate waking life and other people. Without ethics, "awakening" becomes merely another route to ego inflation.

Therefore the book insists: lucid consciousness is measured not by spectacle, but by clarity and compassion.

Science Note: Modern psychology frames "capture" in terms of reward systems, habit loops, and social reinforcement. Ethical development is not reducible to neuroscience, but self-regulation and empathy have measurable correlates and can be cultivated through practices that increase reflection and reduce impulsivity.

VI. Ancient Wisdom in the Dark Recesses

The modern era often caricatures ancient wisdom as superstition. Yet ancient traditions were, in their own domain, empirically oriented: they tested the mind with the mind. They developed repeatable procedures—breath regulation, visualization, mantra, ritual, dream incubation—designed to induce specific states. They preserved results in mythic form because myth is a technology for memory.

What the modern mind calls "dark recesses" the older mind called the underworld, the sea, the womb, the cave. These are not merely places; they are metaphors for the nonconscious layers of psyche where raw emotion, fear, desire, and archetypal imagery reside.

To go inward is to descend. The descent is frightening not because it is evil but because it is unedited. In waking life the person maintains a narrative self. In dreams and trance the editorial voice weakens. The psyche displays itself in symbols.

This is why inner work often begins with discomfort. The mind must encounter what it has ignored. Yet within these depths lie not only wounds but wisdom: intuition, creativity, spiritual insight, and the strange intelligence of the organism.

The teaching insists that wisdom is not always a thought. Sometimes it is an image that arrives with undeniable authority: a door, a staircase, a mirror, a voice that is not the ordinary inner monologue. These encounters can transform a life. Whether they are communications from an astral realm or emergent messages from the deep mind, they function as guidance.

Therefore the book adopts a disciplined openness. It does not dismiss the mystery. It does not demand credulity. It asks the reader to train, record, and observe patterns over time.

Science Note: Much cognition is nonconscious, including emotional learning and implicit memory. Dream content draws heavily from memory fragments and emotional concerns. The idea of "archetypal" imagery is debated in psychology, but cross-cultural motifs do appear. Their interpretation, however, is not a settled science.

VII. The Triangle of Reality and the Circle of Awareness

Imagine an equilateral triangle. Inside it, a circle. The triangle symbolizes the apparent structure of reality—bounded, stable, measurable. The circle symbolizes awareness—the living capacity to know.

The triangle has three faces:

1. **The physical plane:** the body, sensation, objects, movement, time as sequence.

2. **The mental plane:** thought, emotion, imagination, language, interpretation.

3. **The spiritual plane:** meaning, conscience, unity, the intuition of the transcendent.

These are not scientific categories. They are phenomenological categories—ways the human being experiences life. The triangle is a map of lived reality.

In psychic slumber, awareness remains small and clings to one face of the triangle, usually the physical. The person becomes a materialist by habit, not by philosophy. The person assumes the senses are the limit of what is real because the senses are what dominate attention.

Awakening begins when the circle expands. As awareness expands, it recognizes that thought is not the same as fact, that emotion is information but not always truth, that the body is a vehicle but not the whole self, and that meaning is not merely a social construct but a lived dimension.

When the circle expands beyond the triangle, the teaching calls it god-consciousness—not in the sense of personal divinity, but in the sense of unity: awareness that recognizes interconnectedness and is no longer hypnotized by separateness.

This expansion is the foundation for lucid dreaming and astral practice. Without the capacity to observe and choose, the inner traveler remains at the mercy of impulses and symbols. With expanded awareness, the traveler gains the ability to navigate.

Science Note: The "circle expanding" aligns with increased metacognition and self-monitoring. Research on mindfulness and metacognitive training suggests that observing thoughts as events rather than truths can reduce reactivity. The "spiritual plane" here is a meaning-making dimension, which psychology studies indirectly through values, purpose, and self-transcendent experiences.

VIII. Id, Ego, Superego: A Mythic Map of Forces

Freud's model—id, ego, superego—appears in the older text not as a clinical diagnosis but as a mythic map. It describes forces that most people recognize in themselves: appetite, mediator, conscience.

- The **id** wants immediate satisfaction.
- The **superego** wants moral perfection and social conformity.
- The **ego** negotiates, balances, and maintains a workable self.

The esoteric teaching overlays these forces onto the triangle: id as physical impulse, ego as mental mediator, superego as spiritual conscience. The mapping is not literal. It is functional. It provides a language for internal conflict.

Psychic slumber often takes the form of ego servitude: the ego becomes a clerk for impulse, or a servant of social approval, rather than a sovereign agent of awareness. In that servitude the person becomes predictable. The person can be manipulated by desire and fear.

Lucid practice aims to free the ego from automaticity—not to destroy desire, but to place desire within a wider field of choice. In dreams, this means recognizing impulse-driven narratives and choosing differently. In waking life, it means noticing craving before it becomes action.

This is why dream training matters. The dream is a laboratory for impulse. It reveals the mind's reflexes without polite disguise.

Science Note: Freud's structural model is not a direct neuroscientific map, but modern research supports the existence of competing motivational systems and executive control processes. Lucid dreaming appears to involve increased self-reflection during REM sleep, suggesting partial restoration of executive monitoring.

IX. Objective and Subjective Mind: The Modern Threshold

The older teaching describes two minds:

- The **objective mind** operates through the five senses and reasoning. It dominates waking life.
- The **subjective mind** operates through memory, association, and deduction from accepted premises. It

dominates sleep and trance.

In the subjective mode, the mind is less constrained by sensory input. It generates worlds from within. It accepts its own productions with startling confidence. That is why a dream can present impossible architecture, shifting identities, and absurd plots while feeling completely normal.

Lucidity is the moment the objective mind's critical function returns inside the dream. It is the moment the mind says, "This is a dream," and the circle of awareness expands. The dream becomes transparent.

This threshold is the practical target of training. The goal is not to interpret dreams endlessly. The goal is to become conscious within them.

The older teaching also notes that the subjective mind is strongly influenced by suggestion. This is not merely metaphysical; it is pragmatic. The person who trains intention before sleep—through affirmations, visualizations, and ritual—often changes dream content and increases lucidity.

Science Note: Dreaming involves altered executive function and reduced reality testing, especially in REM sleep. Lucid dreaming research includes laboratory verification using prearranged eye movement signals during REM, supporting the claim that reflective awareness can occur within REM dreams (LaBerge, 1985; see also later replications and reviews). Suggestion and intention-setting are consistent with cognitive models of prospective memory and expectation.

X. Concept + Conscious Awareness = Experienced Reality

The older text offers a formula:

Concept + Conscious Awareness = Reality.

In a hybrid interpretation, this becomes:

Interpretation + Attention → Experienced reality.

This does not mean the external world is imaginary. It means experience is mediated. A person does not experience raw data; the person experiences data organized into concepts and highlighted by attention.

In waking life, the senses provide input, concepts label it, and awareness renders it meaningful. In dreaming, the process is similar but the source shifts: memory and emotion supply the input, concepts organize it into narrative, and awareness renders it real—unless lucidity arrives.

Therefore, the training lever is clear: alter attention, alter expectation, alter concept-binding, and the experienced world changes.

This is not an argument for solipsism. It is an argument for responsibility. If attention and concept shape reality as lived, then the person is not merely a passive recipient. The person is a participant.

Lucid dreaming is the most dramatic demonstration of participation. The dreamer changes the world by changing the mind's stance toward it. The waking life also changes—more slowly, but more profoundly—when the person learns to observe and choose.

Science Note: Predictive processing models in cognitive science propose that perception involves top-down predictions constrained by bottom-up input. While the book does not commit to a specific model, the idea that expectations shape experience has substantial empirical support. Dreams can be viewed as internally generated simulations with loosened sensory constraints.

XI. Dreams as Doorways: Partial Transitions

Every night, the human being transitions. The body lies down, sensory input reduces, the nervous system cycles through stages of sleep, and consciousness enters altered forms. In the esoteric language, this is travel. In the scientific language, it is neurophysiology. In lived experience, it is a doorway.

The older teaching calls dreaming a partial transition between planes. The mind loosens its grip on the physical and moves into a world composed of images, memory, emotion, and symbol. The dream is a threshold space: not fully physical, not fully abstract, but experiential.

This threshold is valuable because it is frequent. Most people will never have a dramatic mystical revelation in waking life. Most people will, however, dream thousands of times. The doorway is accessible.

The dream also reveals hidden structure. It exposes the mind's fears, desires, unresolved conflicts, and creative potentials. It shows what the person rehearses unconsciously. It displays the narratives that govern behavior.

Therefore the dream is not a random hallucination to be dismissed. It is a nightly disclosure. If approached with discipline, it becomes a teacher.

Lucid dreaming intensifies this teaching. In lucidity, the dreamer can ask direct questions, test the stability of the environment, and observe how expectation shapes phenomena. The dream becomes a laboratory where metaphysical claims can be explored subjectively and where psychological patterns can be confronted directly.

Science Note: Vivid dreaming is strongly associated with REM sleep but can occur in other stages. Dreams often incorporate recent experiences, emotional themes, and memory fragments. Lucid dreaming is a distinct phenomenon with research support, including verified signaling from within REM sleep (LaBerge, 1985; subsequent work has expanded and refined these findings).

XII. The Astral Plane As Buffer Zone

The astral plane, as described in esoteric traditions, is a buffer zone between physical and mental realities. It is a region of experience where form is plastic, where thought and emotion quickly shape environment, and where symbolic content takes on apparent externality.

The hybrid stance used in this book treats "astral plane" as a name for a family of experiences: hypnagogic imagery, sleep paralysis thresholds, lucid dreams, out-of-body sensations, and the perception of traveling in a nonphysical environment. The term is preserved because it is historically meaningful and experientially descriptive, even if it cannot be confirmed as a separate ontological realm by current scientific methods.

In this buffer zone, the traveler discovers a crucial principle: *inner posture determines outer experience.* Fear tends to summon frightening imagery. Aggression tends to generate conflict. Calm tends to stabilize and clarify. Intention tends to organize the dream toward a goal. This is not moralism; it is mechanics.

Because the astral buffer is responsive, it magnifies the consequences of inner states. It reveals the person's default emotional climate. It also provides a training ground for changing that climate.

The older teaching speaks of layers within the astral. Some feel close to waking life—rooms, streets, familiar faces. Others become abstract or mythic—temples, endless libraries, landscapes that obey symbolic rather than physical law. The traveler should not rush to interpret these layers as literal geography. But the traveler should also not dismiss their coherence. Across many nights, patterns emerge.

The disciplined approach is to record, test, and refine. The person becomes a cartographer of the inner world.

Science Note: Transitional sleep states (hypnagogia and hypnopompia) can produce vivid imagery and sensory distortions. Sleep paralysis can involve intense hallucinations and felt presences. These phenomena are well-documented. Whether they correspond to "planes" beyond the brain is not established; the book uses "plane" as an experiential map.

XIII. Lucid Dreaming and Out-of-Body Experience

Lucid dreaming and out-of-body experience (OBE) are related but distinct. Lucid dreaming is defined by awareness: the dreamer knows the experience is a dream while it is happening. OBE is defined by self-location: the person experiences the self as located outside the physical body, often with a floating or elevated viewpoint.

Some lucid dreams evolve into OBEs. Some OBEs begin from sleep paralysis and then transition into dream-like environments. Some experiences occupy a middle zone where labels become less important than clarity.

The older text tends to treat these experiences as forms of astral projection. The hybrid approach distinguishes them while respecting their overlap. The practical goal is not to argue about taxonomy; it is to cultivate stable, repeatable access to lucidity and to navigate fear, excitement, and expectation.

The most common obstacle is not technique. It is emotional reflex. The moment lucidity arises, excitement spikes and the dream collapses. Or fear spikes and the dream becomes hostile. Therefore training must include emotional regulation and stabilization.

The traveler should also understand that OBEs can be startling and can challenge identity. They can produce profound certainty in the experiencer. Yet certainty is not proof. The wise approach is to hold the experience with respect and skepticism simultaneously: respect for the impact, skepticism about metaphysical conclusions.

Science Note: Lucid dreaming has been demonstrated in sleep labs using eye movement signaling during REM (LaBerge, 1985). Neuroimaging and EEG studies suggest that lucidity involves changes in brain activity compared to non-lucid REM, including increased self-reflective processes (e.g., Voss et al., 2009; Dresler et al., 2012). OBEs are studied in relation to multisensory integration and body representation, with evidence implicating temporoparietal regions in altered self-location (Blanke et al., 2002).

XIV. Astral Light Descending: The First Clear Seeing

The older teaching describes "astral light descending" as fear and misunderstanding being replaced by hope and enlightenment. In lived terms, this "light" is the arrival of clarity within the dream. It is the first time the dreamer sees the dream as dream while still inside it.

This moment is often accompanied by awe. The environment becomes vivid. The dreamer feels a shift in identity, as if a deeper self has awakened. It can also be accompanied by instability, because the mind is balancing between sleeping immersion and waking reflection.

The book treats this moment as sacred and practical. Sacred, because it reveals that consciousness can wake up inside its own projection. Practical, because the moment must be stabilized and trained.

Astral light is not a one-time event. It can become a reliable faculty. Over time the dreamer can recognize recurring dream signs, maintain calm, and sustain lucidity long enough to explore.

When lucidity stabilizes, the traveler begins to notice that the dream environment responds to subtle expectations. A door opens to what is expected behind it. A landscape shifts to match the dreamer's emotional state. A "guide" appears when the dreamer believes guidance is possible. The inner world becomes a mirror.

Hence the title of the book: *The Infinite Mirror.* The inner world reflects the outer, the outer reflects the inner, and both reflect the mind's stance. The mirror is infinite because reflection continues at every scale: cell, person, society, cosmos.

Science Note: Dream control is strongly influenced by expectation and attention. "Vividness" can be influenced by arousal and sleep stage. The experience of "a deeper self waking" can be interpreted psychologically as increased metacognitive awareness during REM.

XV. The Foundation: Recall, Rhythm, and the Dream Journal

Without recall, there is no training. A person may become lucid and forget it by breakfast. The mind's memory systems treat dreams as low-priority unless repeatedly signaled otherwise. The dream journal is the signal.

The Dream Journal as Core Practice

Materials: a notebook and pen by the bed, or a phone note app set to low brightness.

Rule: record immediately upon waking, before movement and before stimulation.

Frequency: daily, even when recall is poor.

Procedure:

1. **Wake gently.** Keep eyes closed. Do not reach for the phone.
2. **Search backward.** Ask: "What was the last image? The last emotion? The last sentence?"
3. **Write fragments first.** Even single words matter: "blue hallway," "panic," "airport," "lost shoes."
4. **Reconstruct.** After fragments are captured, build narrative.
5. **Tag dream signs.** Identify anomalies: changing text, impossible rooms, deceased people alive, flying, repeated locations.
6. **Tag emotional tone.** Fear, awe, desire, shame, calm, curiosity.
7. **Weekly review.** Compile a list of recurring dream signs.

This practice accomplishes several things at once. It improves recall. It increases attention to dreams. It trains pattern recognition. It also builds a relationship with the subjective mind: the mind learns that dream material will be received rather than dismissed.

Sleep Rhythm and the Inner Doorway

Lucid dreaming is facilitated by stable sleep. Erratic sleep fragments REM patterns and reduces dream coherence. Therefore, sleep hygiene is not a moral demand but a technical requirement.

Basic rhythm guidelines:

- consistent sleep and wake times when possible,
- sufficient sleep duration,
- reduced alcohol (which can fragment sleep),
- careful use of caffeine, especially late day.

Lucid practice should not become an excuse for sleep deprivation. Sleep deprivation increases odd phenomena but decreases stability and mental health.

Science Note: Dream recall improves with intention and immediate recording. Sleep fragmentation and substance use can alter REM patterns and dream content. Training prospective memory (remembering to remember) is central to lucid induction techniques.

XVI. Induction: Reliable Gateways to Lucidity

The older teaching uses visualization, affirmation, and autosuggestion. Modern lucid dreaming research and practitioner communities offer structured methods that align well with those tools. This book emphasizes methods with both tradition and evidence: reality testing, MILD, and WBTB.

A. Reality Testing (Daytime)

Reality testing trains the habit of questioning state. If questioning becomes habitual during waking life, it can appear in dreams.

Key principle: the test must be sincere. A mechanical check done mindlessly will not transfer into dreams because it has no attention behind it.

Recommended reality checks:

1. **Memory continuity test:** "How did I get here?" (What

happened 5 minutes ago? 30 minutes ago?)

2. **Text check:** read a line of text, look away, read again. Dream text often shifts.

3. **Time check:** look at a clock, look away, look back. Time often changes oddly in dreams.

4. **Hand check:** examine hands closely; in dreams they may distort.

Daily quota: 5–10 sincere checks, triggered by dream signs discovered in the journal (e.g., whenever encountering mirrors, elevators, classrooms).

B. MILD (Mnemonic Induction of Lucid Dreams)

MILD trains prospective memory: the intention to recognize dreaming.

Before sleep (or after waking during the night):

1. Recall a recent dream.

2. Identify a dream sign.

3. Visualize the dream again, but this time imagine becoming lucid at the dream sign.

4. Repeat with calm conviction:

"Next time I'm dreaming, I will recognize that I'm dreaming."

The sentence is not magic. It is a cue planted in memory, designed to surface during dreaming.

C. WBTB (Wake Back to Bed)

WBTB leverages the natural REM density of later sleep cycles.

Procedure:

1. Set an alarm for ~5–6 hours after sleep onset.

2. Wake for 10–30 minutes. Stay calm, low light.

3. Do MILD during this wake period.

4. Return to sleep with intention.

Many practitioners report that WBTB is the single most reliable amplifier of lucidity. It should be used selectively—2–4 nights per week—so that overall sleep quality is not compromised.

D. Visualization Incubation: The Beach and the Void

The older text uses a "misty void" and a "beach" visualization. These work as incubation scenarios: stable, sensory-rich environments that the mind can enter as it falls asleep.

Beach incubation:

- imagine warm sand, waves, salt smell, distant sun, steady rhythm,
- repeat intention: "Awareness remains as the body sleeps."

Misty void incubation:

- imagine floating in a dark, misty space, calm and weightless,
- imagine awareness as a small light that stays on.

These are not meant to be forced. They are gentle cues.

Science Note: MILD and WBTB have empirical and practical support in lucid dreaming communities and research. Lucidity is not guaranteed; it varies by individual differences and sleep conditions. The mechanisms likely involve prospective memory and increased cortical arousal during REM.

XVII. Stabilization: Extending the Dream and Deepening the Plane

Lucidity is often fragile. The dream collapses into waking or into non-lucid dreaming if the dreamer becomes too excited or too analytical. Stabilization is therefore trained like a skill.

The First 30 Seconds Rule

When lucidity arises, the dreamer should not immediately attempt dramatic feats. The first 30 seconds are used to stabilize.

Stabilization sequence:

1. **Breathe slowly.** Relax the urgency to perform.
2. **Engage touch.** Touch a wall, the ground, or an object; feel texture.
3. **Rub hands.** The sensation anchors attention in the dream body.
4. **Look at details.** Observe fine patterns: wood grain, fabric weave, pebbles.
5. **Speak clarity.** Say aloud: "Clarity now," or "Stabilize."

If the dream begins to fade:

- **spin gently** (a known technique among practitioners),
- or **focus on a single object** until the scene returns,
- or **drop to the ground** and feel it.

Emotional Stabilization

The dream environment often mirrors emotion. Therefore the dreamer should cultivate an inner demeanor: calm curiosity. Curiosity stabilizes. Fear destabilizes. Excitement destabilizes.

A useful stance is: *This is a training session.* That stance reduces pressure and preserves attention.

The Deepening Choice

Once stability is achieved, the dreamer can choose one of two directions:

- **Exploration:** observe the environment, ask questions, meet figures, explore themes.
- **Control training:** practice abilities like flight or transformation to strengthen intention.

Both are valuable. Control training strengthens agency. Exploration strengthens insight.

Science Note: Stabilization techniques likely work by increasing attentional engagement and reducing abrupt arousal shifts that trigger awakening. Speaking and intentional actions within lucid dreams have been reported and studied; they may recruit self-reflective networks that maintain lucidity.

XVIII. Astral Projection Practice: The Deliberate Exit

Astral projection practice, in the esoteric tradition, is deliberate separation: shifting the sense of self from the physical body into a subtle body. In the hybrid model, this can be approached as a method for inducing OBEs and lucid dream transitions.

The practice should be done when the body is rested and when the practitioner can afford sleep time. It is not a contest of willpower.

Preparation

- Use WBTB: wake after 5–6 hours, then return to bed.
- Sleep on the back if comfortable (many report easier transitions, though it can increase sleep paralysis).
- Reduce external noise and light.
- Set a clear intention: "Remain calm; recognize the threshold; transition with clarity."

Step-by-Step: The Deliberate Exit

1. **Progressive relaxation:** release tension from feet upward.
2. **Breath regulation:** slow and even; do not force deep breathing.
3. **Witnessing hypnagogia:** observe images and sensations without chasing.
4. **Threshold signs:** buzzing, vibration, heaviness, floating sensations, auditory shifts.
5. **Shift identity:** imagine "awareness" moving one inch above the body, then returning; repeat gently.
6. **Exit motion:** choose one and commit softly:
   - **Roll out:** imagine rolling sideways out of the body.
   - **Float up:** imagine rising like a balloon.
   - **Rope climb:** imagine climbing an invisible rope hand over hand.

Handling Sleep Paralysis

Sleep paralysis can appear as a threshold. It may include chest pressure, fear, hallucinated presences, or loud sounds. The rule is: do not panic. Panic amplifies imagery.

Technique: focus on breath, relax the jaw and forehead, and repeat a simple phrase: "This is a threshold; remain calm." Then continue the exit motion gently.

After Exit: Stabilize Immediately

If the practitioner experiences separation:

- perform the same stabilization sequence used in lucid dreams (touch, hands, clarity),
- avoid rushing to "prove" the experience,

- set a simple goal: move to a door; observe a room; ask for guidance.

The goal is repeatability, not drama.

Science Note: Sleep paralysis is a REM-related state in which muscle atonia persists into awareness. Hallucinations and sensed presences are common. OBEs may involve shifts in body representation and multisensory integration. The hybrid approach treats "exit" as a phenomenological maneuver that can reorganize self-location.

XIX. The "Abilities": Flight, Phasing, Transformation, Travel

In the esoteric narrative, abilities are natural functions of the subtle body. In the hybrid narrative, they are skills of dream control and symbolic command. Either way, they are trained through expectation, emotion regulation, and narrative mechanics.

The practitioner should treat abilities as exercises for will and clarity, not as proof of metaphysical status.

A. Flight

Flight often fails when the dreamer doubts. Doubt is not intellectual; it is embodied hesitation.

Training progression:

1. **Levitation:** small hops, then hovering a foot above ground.
2. **Glide:** jump and glide like low gravity.
3. **Full flight:** choose a propulsion metaphor: wings, wind, invisible jetpack.

Command: "Lightness now."
Stabilizer: keep eyes on horizon; avoid staring down if fear of

heights appears.

B. Phasing Through Walls

Walls feel solid because waking life trains solidity. The dreamer must replace solidity with permeability.

Technique:

- place palms on the wall,
- imagine the wall as mist or water,
- state: "Permeable,"
- move forward without stopping.

If resistance persists, use a door instead. Dreams like doors.

C. Summoning Objects and People

Trying to materialize something in front of the eyes often fails because it creates performance pressure. Instead, use indirect expectation.

Technique:

- assume the desired item is behind the dreamer,
- turn around and retrieve it,
- or reach into a pocket,
- or open a drawer.

For meeting a person:

- say, "They are around the corner,"
- then walk calmly to the corner.

D. Transformation

Transformation is easier when treated as costume rather than biology.

Technique:

- step behind a curtain, into a closet, or into water,
- decide the new form,
- emerge.

E. Teleportation and Travel

Travel works best with narrative devices: doors, elevators, trains, portals. The dream mind likes transitions.

Technique:

- announce destination clearly,
- use an elevator or door labeled with the destination,
- step through with certainty.

F. Asking the Dream for Truth

A powerful hybrid practice is direct inquiry. The dream can answer with symbol.

Questions:

- "Show the source of this fear."
- "What do I need to understand now?"
- "Show me what I avoid."
- "Show me my next step."

The answer may be unsettling. The practitioner should record without immediate interpretation. Meaning often unfolds over days.

Science Note: Dream control likely depends on expectation and attentional allocation. Indirect summoning methods reduce cognitive conflict. Asking questions may recruit narrative generation that integrates emotional concerns; therapeutic models sometimes use dream work as a window into unresolved issues.

XX. The Return: Integration, Ethics, and God-Consciousness

Lucid practice is not complete when the dream ends. The return is the real test: how the person lives afterward. Without integration, lucid dreaming becomes another form of entertainment and another diversion.

Integration Practices

1. **Immediate journaling:** record the lucid event, the emotional tone, and any insights.
2. **One sentence of meaning:** summarize what the dream taught.
3. **One waking action:** choose a small action aligned with the insight (a conversation, a boundary, a creative step).
4. **Ethical reflection:** ask whether the dream behavior reflected virtues or impulses.

Ethics in the Inner Planes

In lucid dreams the dreamer can dominate characters, manipulate scenarios, and indulge impulses without external consequence. Yet internal consequence remains. The psyche learns from rehearsal. A person who rehearses cruelty, even in dreams, may strengthen cruel reflexes. A person who rehearses courage may strengthen courage.

Therefore, the book recommends a simple ethic: treat dream figures with respect, even if they are symbolic. Speak truthfully. Practice compassion. Choose curiosity over conquest.

God-Consciousness Reframed

God-consciousness is not a badge. It is an expanded circle of awareness that includes:

- the body without obsession,
- the mind without identification,
- the moral dimension without rigidity,
- and the unity of being without grandiosity.

When this awareness grows, the person becomes less reactive. Diversions lose their hypnotic grip. The person can enjoy pleasure without being owned by it, pursue success without sacrificing integrity, and love without using love as a narcotic.

This is the true evidence of awakening: increased freedom and increased responsibility.

Science Note: While "god-consciousness" is not a scientific construct, increased self-regulation, reduced reactivity, and prosocial behavior are measurable outcomes in psychological research. The claim that rehearsal shapes behavior aligns with learning theory and habit formation.

Appendix A — A 30-Day Training Protocol (Hybrid)

This protocol balances effectiveness with sleep health. It assumes the practitioner is an adult without untreated sleep disorders. If insomnia or anxiety worsens, reduce intensity and consult a professional.

Week 1: Recall and Dream Signs

Daily

- Dream journal every morning (minimum: fragments +

emotion).

- Reality testing: 5 sincere checks/day (memory continuity emphasized).
- Evening intention: "I will remember my dreams."

Goal: recall improves; dream sign list begins.

Week 2: Induction Begins

Daily

- Journal.
- Reality testing: 7–10 checks/day, tied to top dream signs.
- MILD every night before sleep (5–10 minutes).

Two nights this week

- WBTB + MILD (wake after 5–6 hours; return to sleep).

Goal: first lucid moment or near-lucid recognition.

Week 3: Stabilization and Control

Daily

- Journal, reality testing.
- MILD nightly.

Three nights this week

- WBTB + MILD.

When lucid

- Apply stabilization sequence.
- Choose one simple goal: "touch the ground and say clarity," "read text twice," or "walk through one door intentionally."

Goal: extend lucidity to 30–60 seconds.

Week 4: Astral Exit and Inquiry

Daily

- Journal, reality testing.
- MILD nightly.

Two nights this week

- WBTB + deliberate exit practice (roll out / float / rope).

When lucid

- Ask one question: "Show me what I need to learn."
- Or practice a single ability.

Goal: repeatable threshold recognition and calmer navigation.

Appendix B — Safety, Sleep Health, And When To Pause

Lucid dreaming practice is generally safe for many people, but it is not universally appropriate.

Pause or reduce training if:

- sleep quality worsens for more than a week,
- nightmares intensify and do not respond to stabilization,
- daytime anxiety increases,
- dissociation or depersonalization worsens,
- sleep paralysis becomes frequent and distressing.

Avoid:

- deliberate sleep deprivation,

- mixing intense lucid practice with heavy alcohol or recreational drugs,
- using lucid practice as a substitute for mental health care.

Consider professional guidance if there is a history of psychosis, bipolar mania, severe trauma symptoms, or untreated sleep disorders.

Science Note: Sleep disruption can affect mood and cognition. Sleep paralysis can be distressing but is typically benign. Mental health conditions can be sensitive to sleep manipulation; caution is warranted.

Appendix C — Science Notes Index (By Topic)

- **Lucid dream verification:** eye-movement signaling in REM sleep (LaBerge, 1985).
- **Neural correlates of lucidity:** EEG/neuroimaging studies suggesting increased self-reflective activity during lucid REM (Voss et al., 2009; Dresler et al., 2012).
- **OBE and self-location:** links to multisensory integration and temporoparietal regions (Blanke et al., 2002).
- **Sleep paralysis:** REM atonia persisting into awareness; hallucinations common (sleep medicine literature).
- **Attention and expectation:** cognitive science supports top-down influences on experience (predictive processing as one framework).

References (selected)

Blanke, O., Ortigue, S., Landis, T., & Seeck, M. (2002). Stimulating illusory own-body perceptions. *Nature, 419*(6904), 269–270.

Dresler, M., Wehrle, R., Spoormaker, V. I., Koch, S. P., Holsboer, F., Steiger, A., ... & Czisch, M. (2012). Neural correlates of dream lucidity obtained from contrasting lucid versus non-lucid REM sleep: A combined EEG/fMRI case study. *Sleep, 35*(7), 1017–1020.

LaBerge, S. (1985). Lucid dreaming: Physiological correlates of consciousness during REM sleep. *The Journal of Mind and Behavior, 6*(2), 251–258.

Voss, U., Holzmann, R., Tuin, I., & Hobson, J. A. (2009). Lucid dreaming: A state of consciousness with features of both waking and non-lucid dreaming. *Sleep, 32*(9), 1191–1200.

CHAPTER THREE

The Paradox of Life and Death

Some Essays Begin as Arguments and End as Confessions.

I. This One Begins as Neither.

It begins as a question that is simple to ask and difficult to live with: How should a person judge a life—especially their own—when outcomes are so often shaped by forces that no one controls? Most of us carry, quietly, a set of assumptions about how the world is supposed to work. We tell ourselves that if we are careful enough, moral enough, smart enough, or spiritually disciplined enough, life will reward those traits with safety, meaning, and coherence. When things go well, the assumption feels confirmed. When things go badly, it turns into an indictment: I must have chosen wrong. Or worse: I must be wrong.

This essay is written against that indictment. It does not deny responsibility. It does not celebrate chaos. It does not claim that intention is all that matters or that consequences do not matter. Instead, it tries to describe a mature position that honors three truths at once: (1) **Choices matter** because they express character and shape the conditions of life. (2) **Outcomes are not fully controlled** because life is an interdependent system and luck is real. (3) **Meaning is still possible** because human beings can cultivate awareness and ethical steadiness even when certainty is unavailable.

The framework used here draws on an ancient philosophy of complementary forces (often summarized as yin and yang), reflections on mortality and the "journey" of a life, and a spiritual vocabulary sometimes called **god-consciousness**: the capacity to act from a higher awareness rather than from fear, vanity, or compulsive ego. These lenses are not offered as competing doctrines. They are offered as ways of seeing that can coexist interdependence without fatalism, agency without omnipotence, reverence without superstition.

Two illustrative lives—Z. and L.—appear in the narrative that follows. They are not presented as case files or moral exemplars. They represent patterns that recur in human experience: the earnest person whose convictions lead them into risk; the conflicted person whose choices are less tidy yet still deeply human; the way an ordinary decision can become a doorway to irreversible consequence; the way meaning can survive even when fairness does not.

If this essay has a purpose beyond explanation, it is to help the reader separate **decision quality** from **outcome quality**. Many of our deepest moral injuries come from confusing the two. We treat tragedy as evidence of failure and good fortune as proof of virtue. That confusion does not make us more responsible; it makes us more cruel—toward ourselves and toward others. The pages ahead argue for a different evaluation: the clarity of intention, the integrity of attention, and the willingness to participate consciously in the cycles of change defining all living things.

II. The Great Mystery of Choice

Human beings hunger for moral arithmetic. We want a life that adds up. We want to believe that if we do good, good will follow; if we plan carefully, danger will be avoided; if we love sincerely, loss will be less likely; if we think clearly, we will not be surprised. Societies reinforce these expectations: laws assign responsibility, schools reward diligence, and moral teachings insist that virtue matters. In many ways, these expectations are necessary. Without them, we would drift into irresponsibility. And yet nearly every adult life eventually encounters the fracture: **the world is not arranged to guarantee that moral effort produces matching outcomes.**

A careful driver can be hit by a reckless one. A kind person can become ill. A courageous person can be punished for speaking the truth. A negligent leader can prosper. A peaceful citizen can be caught in violence they never sought. When people encounter this fracture, they often respond in one of three ways. Some double down on control, insisting that outcomes must mirror desert—if someone suffered, they must have earned it or failed to prevent it. Others collapse into cynicism, concluding that if outcomes are not reliably fair, then responsibility is meaningless. Still others deny the fracture, living as if the old arithmetic still applies, until reality breaks through again. None of these responses is stable. Control becomes cruelty, cynicism becomes emptiness, and denial becomes fragility.

The alternative proposed here is **lucid participation**: acting responsibly while acknowledging that outcomes are shaped by interdependence and luck. This stance does not weaken ethics; it strengthens ethics by placing responsibility where it belongs. It asks not for omnipotence but for integrity.

The language of **yin and yang** helps name the structure of this integrity. Yin–yang is often reduced to stereotypes— light versus dark, masculine versus feminine, good versus bad.

But as a practical philosophy, it is better understood as a grammar of change. Every situation contains complementary forces. Those forces depend on each other to be what they are. Over time, one tends to transform into the other. Balance is not static; it is continuous adjustment. Applied to human life, this view corrects a common mistake: thinking of agency as solitary power. In truth, agency is always braided into conditions—family, culture, economics, biology, institutions, timing, weather, and the unpredictable choices of other people. Yin–yang thinking does not eliminate personal responsibility; it places responsibility inside a more accurate picture of causality. It says: **your choice matters, but your choice is never the only force at work.**

A second lens is the **journey metaphor**. It persists across cultures because it captures something undeniable: a life unfolds through time, and no one sees the whole map at the beginning. People meet forks in the road without knowing what each fork will cost. But the journey metaphor has a dangerous cousin: the belief that the destination validates the journey. People conclude that if the ending is good, the life was good; if the ending is tragic, the life was mistaken; if the outcome is celebrated, the person was wise; if the outcome is humiliating, the person was foolish. This logic creates a hidden tyranny. It makes meaning dependent on survival, reputation, or success —things that cannot be guaranteed and are often distributed unfairly. Used honestly, the journey metaphor teaches the opposite: a life is not a single event; a decision is not a prophecy; an ending is not an essence.

A third lens is **god-consciousness**. The term can trigger resistance because it is often used as either dogma or a promise of protection. Here it means something narrower and more practical: a heightened awareness that loosens the grip of ego and aligns action with deeper values—love, truthfulness, courage, compassion, humility. It is not a claim that the universe will reward you for being awake. It is a claim that wakefulness changes how you choose, how you respond, and how you live with what happens. The ego wants guarantees. The ego wants to be right. The ego wants a story where it is either the hero or the victim. God-consciousness, as used here, is the practice of stepping outside those compulsions. It does not erase fear, grief, or uncertainty. It makes them less likely to govern the whole self.

To show how these lenses operate in lived experience, the essay turns to narrative. Two figures—Z. and L.—are used not as proof but as mirrors. Their stories illustrate a common human pattern: **a choice made under uncertainty can meet consequences no one intended.** That pattern forces an ethical question: how do we judge decisions when the world is not a moral scoreboard? The answer this essay develops is a disciplined distinction between **decision quality** and **outcome quality**, and a commitment to preserve both responsibility and compassion in the space between them.

III. The Element of Fate

People imagine fate as dramatic—thunderclaps, prophecies, obvious warnings. In real life, irreversibility often enters quietly. A person closes a door behind them, thinking about errands. They accept an invitation. They take a road they have taken many times. They decide to stop by someone's house. They step into a crowd. They board a bus. They glance at a clock and choose not to be late. Only afterward does the mind crown the moment with significance, as if the decision itself were glowing with destiny. At the time, it was ordinary.

The ordinariness is part of the terror. It suggests that life is porous: at any moment, you may pass from the familiar into the irrevocable without ceremony. This is where people start demanding certainty from ethics. They want a rule that would have prevented the tragedy. They want to find the "wrong turn" so they can believe they are safer than the one who died. They want the world to be legible.

The stories of Z. and L. begin in this territory, where ordinary decisions meet extraordinary consequences, and where the mind tries—sometimes desperately—to turn grief into logic.

IV. Z.: Conviction, Community, and the Moral Dignity of Showing Up

Z. is remembered, in the way people remember the earnest, as someone made of intention. He read widely, listened closely, and treated ideas as obligations rather than decoration. Some people collect beliefs like furniture. Z. treated belief like citizenship. If he thought something was wrong, he felt pulled—sometimes unwillingly—toward response.

He lived inside community: friends, colleagues, neighbors, arguments that ran late into the night. The world he wanted was not an abstraction; it was a shared life in which people could speak, work, raise children, disagree, forgive, and hope without being crushed by distant machinery of violence or indifference. He did not imagine himself as pure. He did not imagine his side was perfect. But he believed that public life is shaped by who shows up, and that silence, over time, becomes a kind of consent.

When a protest rally is planned—large, public, morally charged—Z. decides to go. He does not treat the decision as heroic. He treats it as consistent. From one angle, the choice is simple: go or don't go. From another angle, it contains an entire metaphysics of human dignity. To show up is to refuse the moral anesthesia of distance. It is to insist that one's body, not just one's opinion, belongs to the world one is criticizing.

In the grammar of yin–yang, Z.'s decision is yang: outward, active, declarative. But it is not yang as aggression. It is yang as participation—the movement of someone who believes conscience must sometimes become visible. This matters ethically. A culture can train people to believe that safety is the highest good, and that the safest life is the most rational life. Yet the pursuit of safety, when made ultimate, becomes an idol. It slowly degrades the capacity to act for anything larger than self-preservation.

Z.'s choice to attend the rally is not proof of moral superiority. It is evidence of a particular orientation: *some risks are worth taking because some values are worth embodying.* Still, values do not command outcomes. To confuse moral intention with control is to set the soul up for either pride or despair.

V. L.: Conflicted Agency and the Burden of Being Human Without Coherence

L.'s story is harder to tell cleanly because L. is harder to summarize. Some people move through life with a consistent narrative about themselves. Others do not. L. belongs to the second group. He is not a villain. He is not a saint. He is a person whose decisions do not line up like soldiers.

L. joins the army. The reasons are mixed in ways familiar to many: pressure, fear, patriotism, economic necessity, a desire to prove something, a desire to belong, a desire to escape a smaller life, or the inability to see other options as real. Even when a person gives one reason out loud, the inner reasons are often plural. People are not single-motive machines. They are weather systems.

At some point, L. goes AWOL. That phrase becomes a stamp in the public mind: irresponsible, cowardly, disloyal. Yet human motives rarely obey stamps. Going AWOL can mean panic. It can mean conscience. It can mean collapse. It can mean refusal. It can mean confusion so thick a person cannot breathe inside it. None of these explanations automatically absolves anyone.

They are not excuses; they are descriptions of the inner terrain in which choices are made.

Here the narrative touches a difficult moral truth: we tend to judge people as if they have perfect access to themselves. Many do not. Many live in partial darkness, with fear and longing tangled together. Their agency is real, yet fractured. Their choices are choices, yet also symptoms. This does not eliminate responsibility; it complicates how responsibility should be assigned. It suggests that moral judgment must be paired with psychological humility.

In yin–yang terms, L. lives inside turbulent oscillation: attempts at assertion followed by retreat; commitment followed by flight; courage followed by dread. This is not a cosmic sentence. It is a portrait of a psyche under strain, pulled between opposing forces and unable to synthesize them into stable balance.

Later, L. decides to visit a friend. The decision has the texture of human need: connection, refuge, conversation, perhaps reconciliation, perhaps distraction. There is nothing inherently dramatic about it. And yet, as with Z., the ordinary becomes a threshold.

VI. Tragedy and the Instinct to Rewrite Causality

Both Z. and L. die as a result of circumstances connected to these decisions. The structural pattern matters: a choice becomes linked to an outcome no one intended, and the mind rushes to make sense of the linkage. After tragedy, people often do one of two things. They moralize the outcome ("If he hadn't gone, he would be alive"), or they mythologize the choice ("He was destined to die that way"). Moralizing preserves the illusion of control. Mythologizing preserves the illusion of meaning. Both can be coping strategies. Both can also become forms of disrespect, reducing a life to a lesson for the living.

A more truthful response is harder: to hold that an outcome can be devastating without being a verdict, and meaningful without being a plan. This is where the distinction between **decision quality** and **outcome quality** becomes ethically urgent.

Outcome quality asks: did it end well? Decision quality asks: was it chosen with integrity given what could reasonably be known? The temptation is to let the outcome retroactively colonize the decision. If Z. died, then going was wrong. If L. died, then visiting was foolish. But that is not reasoning. It is grief trying to become logic.

VII. The Ethics of Uncertainty: Responsibility Without Guarantees

To say that outcomes involve luck is not to say choices are irrelevant. It is to place responsibility where it belongs: in what an agent could govern. A responsible person can seek information and avoid obvious dangers, consider how actions affect others, notice when emotion distorts judgment, admit uncertainty, and choose in alignment with values. A responsible person cannot control every other person's actions, prevent every random mechanical failure, foresee every violent eruption in a crowd, guarantee that a friend's home is safe on a particular night, or ensure that timing and chance will not collide.

Responsibility, then, is not omnipotence. It is careful participation. This is a difficult ethic because it denies the fantasy that the moral life is a technique for avoiding pain. It says instead: the moral life is a way of meeting reality with integrity, including when reality injures you.

VIII. Yin–Yang Becomes Personal: Interdependence Inside the Self

Yin–yang is not only about seasons and weather. It also describes the inner world. Within a single person, there are complementary forces: the desire to act and the desire to rest; the impulse to speak and the impulse to remain silent; the need for safety and the need for meaning; the wish to belong and the wish to be free; the hunger for certainty and the awareness that certainty is not available.

Wisdom is not choosing one side forever. Wisdom is noticing the shifting pattern and responding with sensitivity to context. Z.'s life leans toward engagement, public action, declaration. L.'s life leans toward rupture and retreat, attempt and collapse. Neither pattern is automatically virtuous or vicious. The ethical question is whether the person is becoming more awake inside the pattern—more truthful, more compassionate, more integrated—or whether they are becoming more reactive and trapped.

IX. God-consciousness as Integration: Reason, Emotion, and the Ego's Demands

People often speak as if reason and emotion are enemies. This misunderstanding creates needless suffering. Emotion is not the opposite of intelligence; it is information—sometimes accurate, sometimes distorted. Reason is not the opposite of humanity; it is a tool for clarifying consequences, comparing values, and resisting impulse.

The balance implied by god-consciousness is a practical wisdom: use reason to map likely consequences and test assumptions; use emotion to detect what matters, where harm is happening, and what values are alive; use awareness to notice when either is hijacked by ego. In this sense, god-consciousness is not a mystical escape from psychology. It is intensified honesty about psychology.

The ego wants guarantees. It wants to be right. It wants a story where it is either hero or victim. Awareness loosens these compulsions. A person acting from god-consciousness may still feel fear, but fear is less likely to become the sole author of the decision. They may still feel pride, but pride is less likely to demand a world that applauds them. They may still feel anger, but anger is less likely to become cruelty. None of this immunizes anyone from tragedy. It changes how one travels through it.

X. Mortality and the Refusal to Make Death the Measure of Life

The deaths of Z. and L. force an encounter with mortality. Not mortality as abstraction, but mortality as rupture that reorders the living. When someone dies, the living often rearrange the past in an attempt to make the death comprehensible: "What was the point?" "What should have been different?" "Who is to blame?" "What lesson is being taught?" Sometimes there is blame; sometimes there are lessons. But sometimes the deepest truth is simpler: life is vulnerable by design.

The cycle of seasons offers a language for this vulnerability. Seeds break open in darkness. Trees lose leaves. The earth appears dead and then returns to green. Nothing living avoids change; nothing living avoids endings. To say this is not to romanticize loss. It is to resist an additional violence: the demand that death must be deserved, or meaningful in a way that makes it acceptable. A person can honor the dead without explaining them away.

XI. The Egg: Potential, Rupture, and the Ethics of Transformation

The symbol of the Egg appears because it captures something psychologically exact: transformation often requires the breaking of an enclosure. Inside the egg is potential. Outside the egg is a world that will injure and nourish. The shell is both protection and limitation. To be born is to lose safety. To grow is to accept exposure.

Ethically, the Egg symbolizes a demand placed on every person: to become more than the defensive self. Many people live their whole lives inside shells—identities, habits, resentments, certainties—because shells feel safe. But shells also suffocate. God-consciousness, in this symbolic language, is a willingness to let the shell crack: to let pride, fear, and self-protection loosen enough for love and clarity to breathe.

Z., by stepping into public risk, enacts one kind of cracking—exposure chosen for the sake of value. L., by moving toward connection despite inner turbulence, enacts another—exposure chosen for the sake of being human with others rather than alone with fear. The outcomes are tragic. Yet the ethical movement can still be recognized: the attempt to live as more than a frightened organism guarding itself.

XII. What We Owe the Living: Beyond Sogans After Tragedy

After tragedy, people speak in absolutes: "Never do that again." "Always follow your conscience." "Protests are dangerous; stay home." "The world is meaningless; protect yourself." "It was his own fault." Each absolute tries to end uncertainty. Each tries to close the wound with a slogan. But a mature response looks less tidy. It accepts that risk cannot be eliminated without shrinking life; that some risks are unjust and should be resisted collectively; that individuals have different thresholds and responsibilities; that grief will make the past feel like a courtroom; that the mind searches for reasons even when the world offers only conditions.

What we owe the living is not false certainty. What we owe is a more humane way to stand in uncertainty without becoming reckless or paralyzed.

XIII. The Moral Center: How to Judge a Decision When the World is Not a Scoreboard

If the narrative teaches anything coherent, it is a way to evaluate decisions without collapsing into fatalism or control-moralism. A decision can be judged by asking: What did the person reasonably know—not what we know now, but what was available then? What did the person intend—not what the outcome suggests, but what they aimed at? What values guided the choice—fear, vanity, compassion, justice, belonging, love, duty, escape? What alternatives were realistically open—live options given constraints? How did the person hold risk—deny it, inflate it, accept it, mitigate it, share it? And how did the person treat others in the process—did they respect other people's vulnerability and agency?

This approach permits moral evaluation without pretending outcomes are proof. It also leaves room for learning. A person can say: "My decision was understandable but still unwise," or "My decision was sound, and tragedy still happened." Both statements can be true. The human heart resists layered truths because it wants a single verdict. But maturity often means living with complexity without using it as an excuse.

XVI. The Final Paradox: Acting Fully Without Demanding Guarantees

The paradox at the heart of this essay is simple to state and difficult to practice: act as if your choices matter—because they do; release the demand that your choices control outcomes—because they do not. Yin–yang offers a language for it: action and receptivity, assertion and humility, initiative and acceptance. God-consciousness offers a discipline for it: ego loosening, awareness deepening, values clarifying, love becoming less conditional on success.

Z. and L. show what happens when this paradox is lived inside real conditions. Sometimes courage meets tragedy. Sometimes confusion meets tragedy. Sometimes life ends without offering an explanation that satisfies the living. And still, we choose. Meaning is not the guarantee of safety. Meaning is the willingness to live awake in a world where safety is never absolute.

BIBLIOGRAPHY

Bremmer, Jan N. *Initiation into the Mysteries of the Ancient World*. Berlin: De Gruyter, 2014.

Burkert, Walter. *Ancient Mystery Cults*. Cambridge, MA: Harvard University Press, 1987.

Eliade, Mircea. *Myth and Reality*. Translated by Willard R. Trask. New York: Harper & Row, 1963.

Eliade, Mircea. *The Myth of the Eternal Return: Cosmos and History*. Translated by Willard R. Trask. Princeton, NJ: Princeton University Press, 1954.

Graham, A. C. *Disputers of the Tao: Philosophical Argument in Ancient China*. La Salle, IL: Open Court, 1989.

Graf, Fritz. *Eleusis und die orphische Dichtung Athens in vorhellenistischer Zeit*. Berlin: de Gruyter, 1974.

James, William. *The Varieties of Religious Experience: A Study in Human Nature*. New York: Longmans, Green, 1902.

Kahneman, Daniel. *Thinking, Fast and Slow*. New York: Farrar, Straus and Giroux, 2011.

Nagel, Thomas. *Mortal Questions*. Cambridge: Cambridge University Press, 1979.

Schwartz, Benjamin I. *The World of Thought in Ancient China*. Cambridge, MA: Harvard University Press, 1985.

Slingerland, Edward. *Trying Not to Try: Ancient China, Modern Science, and the Power of Spontaneity*. New York: Crown, 2014.

Wang, Robin R. *Yinyang: The Way of Heaven and Earth in Chinese Thought and Culture*. Cambridge: Cambridge University Press, 2012.

Williams, Bernard. *Moral Luck: Philosophical Papers 1973–1980*. Cambridge: Cambridge University Press, 1981.

CHAPTER FOUR

Symbolism of the Mind Graph

Throughout history, man has sought to know what lies beyond the veil of death.

I. The Great Mystery is Revealed to Us

Books have been written; folklore, myths, and legends have been told. This mystery has haunted the imagination through the ages.

The veil of death represents the unknown realm beyond mortal existence—a threshold separating the known world from what may follow. The desire to understand what happens after death is a fundamental aspect of human curiosity, driving philosophical inquiry, religious belief, and artistic expression.

Many cultures preserved sacred texts exploring the afterlife and the journey of the soul: the *Egyptian Book of the Dead*, the *Tibetan Book of the Dead*, and scriptures such as the Bible, the Quran, and the *Bhagavad Gita*. Myths and legends likewise describe the afterlife, deities, and supernatural realms—Greek accounts of the Underworld, Norse Valhalla, and the Celtic Otherworld—

reflecting humanity's attempt to grapple with death and what lies beyond.

The Great Mystery has inspired countless works of art, literature, and music. From ancient epics to modern novels, the theme of life after death continues to captivate and challenge the imagination.

Contemplating death and the afterlife is also profoundly personal. It provokes existential questions about life's meaning, the nature of the soul, and the possibility of immortality.

Ancient Egyptians imagined an elaborate afterlife in which the soul journeyed through the Duat and was judged by Osiris; the worthy entered the Field of Reeds. Greeks and Romans described realms such as Hades, the Elysian Fields, and Tartarus, where souls were sent according to how they lived. Hindu and Buddhist traditions emphasize the cycle of reincarnation (samsara) and the ultimate goal of liberation (moksha or nirvana), treating the afterlife as continuation of the soul's movement toward enlightenment. Taoist views describe the soul's passage into the afterworld, seeking harmony with the Tao and, through spiritual cultivation, immortality.

Modern inquiry has explored near-death experiences, in which individuals report visions and sensations suggesting—though not proving—continuation of consciousness after death. The study of paranormal phenomena, including ghosts and spirits, likewise attempts to provide evidence of life after death. Philosophers such as Søren Kierkegaard, Martin Heidegger, and Jean-Paul Sartre explored how death shapes human existence, freedom, and authenticity. Meditation, prayer, and other spiritual disciplines aim to connect individuals with higher states of consciousness and the divine, offering personal experiences that some interpret as glimpses of the afterlife's nature.

The Great Mystery of what lies beyond death remains central

in human thought and culture. The veil of death continues to inspire curiosity and contemplation. By engaging diverse perspectives—sacred texts, myth, philosophy, spiritual practice, and scientific inquiry—we can better understand why this mystery matters and how it shapes our lives, fostering connection to the eternal and the transcendent.

Reflecting on these themes invites us to explore our own beliefs and experiences with openness and curiosity.

II. I Think, Therefore I Am

Thus far, we have assumed that we exist and that reality is our perceived physical condition. We accept that this existence is actual. We also understand that the human brain converts sensory data into impulses, and that the mind translates those impulses into meaningful concepts. These ideas—together with the memories associated with them—create the fabric of our reality.

The foundational assumption is that we exist, a premise often associated with René Descartes' "Cogito, ergo sum" ("I think, therefore I am"). We understand reality as the physical world we perceive through the senses, and we generally treat that perceived reality as actual and objective.

The brain converts sensory input—sight, sound, smell, taste, and touch—into neural impulses. The mind interprets these impulses through complex cognitive processes: perception, interpretation, and integration. From this process, we form ideas—our understandings of the world.

Memories are stored experiences and knowledge that provide context and continuity. They help us navigate and make sense of what we perceive. Yet while we assume an objective reality, our experience of it is inherently subjective. Each person's reality is shaped by how their sensory data is processed, by cognition, and by memory. Still, people share enough common perceptions and concepts to communicate and interact within a collectively agreed-upon reality.

Phenomenology focuses on the structures of experience and consciousness, emphasizing the subjective nature of perception; Husserl argued that understanding the essence of experience is critical to understanding reality. Empiricism holds that expediencies primarily from sensory experience; Locke and Hume emphasized the role of the senses in forming ideas, and Locke's "tabula rasa" suggests knowledge derives from experience.

Neuroscience investigates how the brain processes sensory data to form perception—revealing, through illusions and perceptual errors, how easily the brain can be misled and how constructed experience can be. It also studies how the brain encodes, consolidates, and retrieves memories through neural networks. Cognitive psychology explores perception, problem-solving, and concept formation, clarifying mechanisms behind our experience of reality.

By examining these assumptions philosophically and neuroscientifically, we gain a deeper understanding of existence and perception, and of the interplay between sensory data and cognition that constructs the world as we experience it.

III. Sensory Input and Memories

Since reality results from sensory data received through sense receptors, sensory perception directly affects the scope and accuracy of what we believe is real. Sense receptors are limited by design and can be affected by physical or medical conditions. And because memory is based on experience, reality must differ from one individual to the next.

Sense receptors gather environmental data which the brain

processes into perception. The range and limits of each sense shape what can be known. Humans cannot see ultraviolet light, for example, though some animals can; this alone implies different realities. The eye detects light only within a narrow range of wavelengths (approximately 380–750 nanometers). Human hearing typically does not register sounds above 20,000 Hz.

Physical conditions such as blindness or deafness can significantly alter perception and thus the lived experience of reality. Color vision deficiencies change how colors appear; anosmia eliminates olfactory input. Neurological and psychiatric conditions can also affect processing—hallucinations in schizophrenia are one example of perceived stimuli without external sources.

Memories form from experience, but they are not perfect recordings. They are vulnerable to distortion and bias, and recall is influenced by emotion, belief, and later experience. Because perception and memory are subjective, each person constructs a unique reality shaped not only by biology but also by culture, upbringing, and environment.

Despite these differences, people build shared or consensus reality through language, social norms, culture, and science. Shared terms and agreements allow collective navigation of the world. Yet differences in perception and experience can produce misunderstanding. Empathy and the effort to understand another's perspective are essential for bridging these gaps.

Phenomenology emphasizes subjective experience; philosophical relativism holds that viewpoints have no absolute validity outside the contexts from which they arise. Neuroscience and psychology explain why individuals can experience the same stimulus differently, given variability in brains, perception, and memory.

Recognizing the diversity of constructed realities helps us

appreciate the richness of human experience and encourages better communication and cooperation.

IV. Perception is Subjective

Though we may agree that a rose is red and smells sweet, the degree to which one sees red or experiences "sweetness" may differ significantly. We share basic terms like "red" and "sweet," which makes communication possible, but these terms do not guarantee identical perception.

Color perception varies due to genetic differences, cone-cell distributions, and sensory thresholds. Individuals with color vision deficiencies may experience a red rose as a different hue. Culture can also influence how colors are categorized, which can subtly shape perception.

Smell is equally subjective. Genetic variation affects olfactory receptors, changing what is detected and how it is interpreted. Personal experience and memory shape response: a fragrance that feels pleasant to one person may evoke discomfort or neutrality in another because of past associations.

These variations can create misunderstandings when people assume their sensory experiences are universal. Recognizing subjectivity supports empathy and strengthens relationships. Vision science and psychophysics help quantify how stimuli relate to perception; olfaction research examines the genetic basis of odor detection and how the brain processes scent. Phenomenology highlights subjective experience, while intersubjectivity describes how shared reality emerges through negotiation among individual perspectives.

Exploring sensory subjectivity deepens our understanding of how uniquely each person experiences the world and why empathy matters in maintaining a shared reality.

V. Neurochemistry and Brain Environment

To understand these relationships more clearly, we can examine how sensory perception, memory, and emotion interact in the brain.

The hippocampus is crucial for forming and retrieving declarative memories (facts and events) and for consolidating short-term memories into long-term storage. It also supports spatial memory, aiding navigation and the understanding of spatial relationships.

The amygdala processes emotion—especially fear and pleasure—and assigns emotional significance to experience. When an event is emotionally significant, the amygdala "tags" it, increasing the likelihood it will be remembered. Stress hormones such as cortisol can strengthen memory consolidation in acute situations, though chronic stress can impair memory and cognition. Emotional context also influences later recall: positive or negative feelings associated with an event can affect clarity and retrieval.

The prefrontal cortex supports complex cognition, decision-making, and social moderation. It helps regulate emotion and evaluates risk and reward. The basal ganglia contribute to procedural memory—skills and habits—and help control voluntary movement and routine behavior.

Neurochemistry also shapes memory and emotion. Dopamine supports reward-motivated behavior and reinforces positive learning. Serotonin influences mood regulation and affects how memories are formed and recalled. Cortisol modulates memory formation in stress, strengthening or weakening encoding depending on intensity and duration.

Memory formation begins with sensory receptors gathering environmental data. During encoding, sensory information is integrated with existing knowledge and emotional context, engaging multiple brain regions, including the hippocampus and amygdala. Consolidation stabilizes memories over time —often during sleep—through transfer and integration across cortical regions. During retrieval, neural patterns are reactivated, and memories can be updated through reconsolidation, making memory dynamic rather than fixed.

Positive emotional memories can support resilience and well-being; negative emotional memories can contribute to stress, anxiety, and in some cases PTSD. Emotional regulation strategies such as cognitive reappraisal, mindfulness, and therapeutic intervention can help process and integrate emotional memories.

Understanding these mechanisms clarifies how perception, emotion, and memory cooperate to shape experience and the felt reality of our lives.

VI. Brain Evolution and Genetics

The brain is a complex organ of hundreds of billions of neurons organized into intricate neural networks that form the foundation of daily experience. Brain activity is shaped by environment, genetics, culture, emotion, development, and experience. The brain is also highly dynamic, continually changing and creating new channels for information flow.

Neurons communicate through synapses, and each neuron can form thousands of connections. These pathways support cognition, perception, memory, and decision-making. Environmental stimuli provide sensory input, and through neuroplasticity the brain adapts to change by forming and reorganizing synaptic connections in response to learning and experience.

Genetics influences brain structure and function and can affect cognitive capacities and susceptibility to disorders. Environment interacts with genetic predispositions, influencing gene expression and brain activity.

Culture shapes perception and response to the world, affecting cognitive and emotional processes. Language, as a cultural system, influences thought patterns and memory. Emotion affects attention, learning, and recall; emotional states shape cognitive performance.

Brain development continues from infancy through adulthood with critical periods for certain skills, and disruptions can contribute to neurodevelopmental conditions such as autism or ADHD. Experience continually reshapes neural networks —reinforcing some pathways while pruning others. Synaptic plasticity underlies learning; neurogenesis, especially in the hippocampus, supports adaptability; structural changes such as dendrite and axon growth increase connectivity.

Learning new skills can measurably alter brain structure and function. After injury, neuroplasticity enables compensation through reorganization and the formation of new connections.

Understanding these dynamics underscores the brain's capacity for learning, recovery, and adaptation and highlights the importance of environments and practices that support brain health.

VII. Higher Level Thinking

The cerebrum—the most evolved region of the brain—is responsible for higher-level thinking and decision-making. It is divided into two hemispheres connected by a neural pathway that allows information to travel between them. The cerebrum consists of four lobes, each with specialized functions, and effective communication among these lobes and the lower brain centers is essential.

The left and right hemispheres specialize in different cognitive tasks and control the opposite sides of the body. The corpus callosum connects them, enabling coordination.

The frontal lobe supports executive function: planning, problem-solving, decision-making, and regulation of behavior and emotion. It contains the primary motor cortex for voluntary movement. The parietal lobe processes bodily sensation and supports spatial orientation and body awareness; it contains the primary somatosensory cortex. The temporal lobe processes auditory information and is involved in memory, emotion, and language comprehension; it contains the primary auditory cortex. The occipital lobe is dedicated primarily to visual processing and contains the primary visual cortex.

Cerebral functioning depends on communication within and beyond the cerebrum. Association, commissural, and projection fibers connect regions of the cortex and link it to lower centers. The thalamus relays sensory and motor signals to the cerebrum and contributes to consciousness and alertness. The brainstem connects cerebrum and spinal cord and regulates vital functions such as breathing and sleep-wake cycles. The basal ganglia interact extensively with the frontal lobe to coordinate movement and aspects of cognition.

The limbic system—including the hippocampus and amygdala—communicates with the cerebrum to shape emotion and memory. The prefrontal cortex helps regulate emotion by interacting with limbic structures.

These interconnected systems illustrate how the cerebrum integrates sensation, movement, memory, emotion, and reasoning to shape behavior and experience.

VIII. Neuronal Communication and Neural Circuitry

Communication in the brain occurs through an electrochemical process that transmits information from one neuron to the next synapses. A neuron can have up to one hundred thousand synapses, each receiving input from many others. Neural circuitry changes in response to demand, forming new pathways during learning and integrating new information with existing knowledge. This creation and strengthening of neural networks constitutes learning.

Neuronal communication begins with an action potential traveling along the axon. When it reaches the axon terminal, neurotransmitters are released into the synaptic cleft. These chemical messengers bind to receptors on the receiving neuron, producing a response.

The receiving neuron integrates signals through temporal and spatial summation, and synaptic plasticity strengthens or weakens connections over time. Synaptogenesis creates new synapses in response to learning and experience. Hebbian learning—"cells that fire together wire together"—describes how repeated co-activation strengthens a connection. Associative learning links new information to existing knowledge, integrating it into larger networks that support complex cognition.

Long-term potentiation (LTP) is a durable increase in synaptic strength and is a key mechanism of learning and memory. Neurogenesis, especially in the hippocampus, adds new neurons to circuitry and supports learning. Sleep contributes to consolidation as activity patterns linked to learning are replayed, strengthening synaptic connections.

These mechanisms reveal the brain's dynamic ability to communicate and adapt, emphasizing the importance of sustained engagement and stimulation for cognitive health.

IX. Anatomy of the Brain

The mid-brain, located deep within the cerebrum, is crucial in regulating emotion, sleep, sexuality, smell, attention, body regulation, and hormones. Because incoming sensory information passes through this area, the mid-brain plays a key role in directing where signals are routed.

The mid-brain (mesencephalon) sits above the brainstem and below the thalamus and cerebral cortex. It includes structures such as the tectum and tegmentum, and nuclei including the substantia nigra and red nucleus.

It interacts with the limbic system to regulate emotional responses. Dopamine pathways from the substantia nigra and ventral tegmental area influence mood and reward. Through the reticular activating system, the mid-brain supports wakefulness and alertness and contributes to REM sleep, associated with vivid dreaming.

The mid-brain communicates with the hypothalamus, which regulates sexual behavior and hormone release. It integrates olfactory input with other sensory information to influence behavior and memory. The superior colliculus supports visual attention and coordinated eye movement; the reticular formation contributes to attention and arousal.

Through connections with the brainstem and hypothalamus, the mid-brain contributes to autonomic regulation such as heart rate, respiration, and blood pressure. It works with the thalamus in relaying sensory information to appropriate cortical regions, supporting coherent perception. The substantia nigra, a critical component of the basal ganglia, contributes to movement regulation; the mid-brain also communicates with the cerebellum to fine-tune motor activity.

Understanding these functions clarifies the mid-brain's role in routing information and maintaining coordinated brain function and behavior.

X. Autonomic Nervous System

The brain stem is the brain's deepest, oldest, and most primitive part. It controls essential instincts and is the first to respond to trouble, initiating the fight-or-flight response.

The brain stem is located at the base of the brain and connects the cerebrum with the spinal cord. It includes the midbrain, pons, and medulla oblongata.

It regulates life-sustaining functions such as heart rate, breathing, and blood pressure, and it governs reflexes essential for survival, including swallowing, coughing, and sneezing. It is central to the autonomic nervous system, especially the sympathetic system that activates in response to threat. When danger is perceived, it mobilizes the body by increasing heart rate, redirecting blood flow toward large muscles, heightening alertness, and signaling hormone release—adrenaline and cortisol—preparing the body to confront or flee.

The pons serves as a bridge connecting regions such as the cerebellum and cerebral cortex and helps regulate breathing rhythm. The medulla oblongata contains centers controlling heart rate, blood pressure, respiration, and reflexes such as vomiting and sneezing.

The brain stem maintains homeostasis and governs instinctual survival behavior. Its role in routine physiological regulation and emergency response underscores its fundamental importance in survival.

XI. Neuroplasticity and Synaptogenesis

The physical brain is shaped by experience and remains dynamic throughout life. This capacity—neuroplasticity—allows continuous adaptation and growth. With proper enrichment, cognitive skills can be enhanced, and intelligence may be increased without fixed limitation.

Neuroplasticity is the brain's ability to reorganize by forming new neural connections. Synaptogenesis creates new synapses in response to experience. Neurogenesis generates new neurons, especially in the hippocampus. Hebbian learning strengthens pathways through repeated co-activation, and long-term potentiation (LTP) increases synaptic strength following repeated stimulation.

Novel and challenging environments can increase synaptic density and improve cognitive function. Activities that challenge the brain—puzzles, learning new skills, problem-solving—promote plasticity. Although early development contains sensitive periods, adults continue to develop new skills and adapt, demonstrating lifelong plasticity.

Neuroplasticity also supports recovery after injury by enabling reorganization and compensation. Rehabilitation can leverage this capacity to restore function.

Enrichment includes continuous education, learning multiple languages, physical activity, and social engagement. Exercise supports neurogenesis and brain health, and social connection contributes to cognitive and emotional well-being.

Understanding neuroplasticity highlights the potential for lifelong learning and underscores the value of an active, stimulating environment for maintaining and improving brain health.

XII. Constant Learning Reinforces Neural Pathways

Neural pathways created by learning degenerate when they are no longer used, while frequently used pathways become stronger and more permanent. The brain eliminates weak connections through synaptic pruning, streamlining networks and improving efficiency. Repeated use strengthens pathways through mechanisms such as long-term potentiation.

When learning a new activity, the brain initially responds slowly as it forms new connections between neurons, including growth of dendrites and axons. Early learning requires significant cognitive effort. Repeated practice reinforces the relevant pathways, making them more robust. With continued practice, axons become myelinated, improving the speed and efficiency of signal transmission. As a result, information transfers more rapidly and reliably.

Repeated activation leads to synaptic potentiation, improving signal transmission. Hebbian plasticity explains strengthening through co-activation. As pathways become efficient, performance improves and tasks can become automatic, reducing conscious effort and freeing cognitive resources for higher-level functions.

Motor skill learning strengthens pathways in the motor cortex and cerebellum; cognitive skill learning strengthens pathways in regions supporting memory, language, and problem-solving. Repeated behaviors become habits as their pathways are reinforced. Reducing use of pathways linked to undesirable behaviors can weaken them, making replacement easier.

These mechanisms show why practice and consistent engagement are essential to skill acquisition, habit formation, and long-term cognitive maintenance.

XIII. Identifying Patterns

The brain elicits patterns of meaning from the information it receives. Because memories are not stored intact, the brain associates incoming information with experience. Emotions are entwined in neural connections and are crucial to storing and retrieving memories. Associations to past learning experiences enhance learning and increase the efficiency of neural networks.

The brain identifies patterns in sensory input and integrates information across modalities. It matches new information to existing knowledge to support understanding. Memory is stored in fragments across regions; recall reconstructs memories by assembling fragments and often filling gaps through inference and association.

New information is encoded more effectively when linked to what is already known. Context and emotional cues shape encoding and retrieval. The amygdala helps prioritize emotionally significant events, and emotional arousal strengthens consolidation. Mood-congruent memory describes the tendency to retrieve memories consistent with current emotional state. Highly emotional events can produce vivid "flashbulb" memories due to strong association.

Learning improves when it builds on prior knowledge—scaffolding—and when new material is made meaningful through connection to experience. Emotional engagement can strengthen retention. Imagery and mnemonic devices also leverage associations to facilitate recall.

Understanding these mechanisms clarifies why context, association, and emotion are central to learning and memory.

XIV. Conscious Mind vs Unconscious Mind

Consciousness is complex and can be described in several ways. It may be understood as self-awareness, a sense of identity, or the capacity to elicit meaning from experience.

Consciousness includes awareness of thoughts, emotions, and experiences, and it supports a continuous sense of personal identity through time. It allows us to interpret and integrate experience into a coherent narrative.

The conscious mind operates within working memory, which can hold limited information at once—often described as 5 to 9 chunks. Because capacity is limited, the conscious mind focuses attention by filtering irrelevant data to manage cognitive load.

The unconscious mind processes vast amounts of information outside awareness. It performs pattern recognition, language processing, sensory integration, and regulates bodily functions such as heart rate and digestion. Many routine behaviors become unconscious once habitual, such as driving a familiar route or typing.

The unconscious mind can handle multiple streams of information simultaneously and produce quick, intuitive judgments based on experience. It also shapes emotional responses and instincts. Learning without conscious awareness —such as early language acquisition—demonstrates the unconscious mind's intelligence.

Conscious reflection is crucial for novel situations and deliberate action. Together, conscious and unconscious processes enable adaptive behavior: intuitive insight emerges from unconscious processing, and the conscious mind evaluates and chooses how to act. Practices such as mindfulness and meditation can increase awareness of unconscious processes, supporting insight and emotional regulation.

XV. Perception, Analysis, Decision, Action

Graphically, the conscious mind—which is chronological and sequenced—may be viewed as linear (logical). The unconscious mind—dealing in abstracts and probabilities—may be viewed as non-linear (emotional).

The conscious mind processes information sequentially. It supports deliberate reasoning, planning, and structured decision-making. The unconscious mind processes information holistically and probabilistically, integrating diverse inputs at once. It excels at pattern recognition, rapid judgments, and complex processing without conscious deliberation.

Conscious Mind (Linear)

- Step 1: Perception
- Step 2: Analysis
- Step 3: Decision
- Step 4: Action

Unconscious Mind (Non-Linear)

- Emotions
- Patterns
- Memories
- Instincts
- Intuition
- Abstract Concepts

Explanation of the Graphic Representation

Conscious Mind (Linear)

- Step 1: Perception — receiving sensory input.
- Step 2: Analysis — processing and evaluating the information.
- Step 3: Decision — selecting a course of action.
- Step 4: Action — carrying out the choice.

Unconscious Mind (Non-Linear)

- Interconnected elements such as emotions, patterns, memories, instincts, intuition, and abstract concepts.

Together, the linear conscious mind and the non-linear unconscious mind complement each other, enabling comprehensive processing, decision-making, and adaptive behavior.

XVI. Symbolism of the Mind Graph

Symbolically, the mind may be viewed as a three-dimensional matrix—a data cube containing knowledge, wisdom, concepts, memories, experience, and abstract information. The surface of the matrix represents the linear conscious mind. Beneath the surface lies the non-linear unconscious mind. As new data reaches the surface, neural pathways open to receive it, passing it down into memory and assimilating it into the deeper layers of the unconscious.

In this symbolic model, each "cell" of the cube represents a unit of information, and the cube's vast number of cells reflects the mind's immense capacity. The outer layer corresponds to conscious processing—sequential, logical, and surface-level awareness. The inner layers correspond to unconscious processing—abstract, emotional, and probabilistic.

New information contacts the surface, activates pathways, and is integrated. Over time it is absorbed into deeper layers, becoming part of the unconscious patterns that influence perception, memory, and behavior.

Explanation of the Graphic Representation

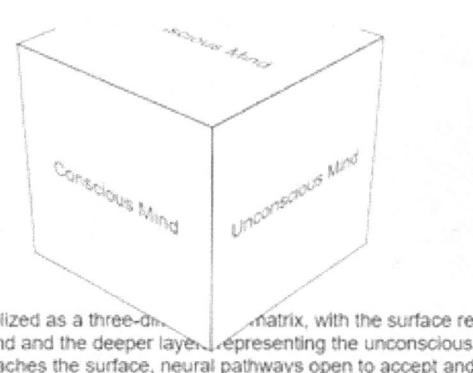

The mind is symbolized as a three-dimensional matrix, with the surface representing the conscious mind and the deeper layers representing the unconscious mind. As new information reaches the surface, neural pathways open to accept and assimilate it into the unconscious mind, forming lasting memories and influencing behavior.

The cube's front and top faces represent the conscious mind —the surface where linear processing occurs. The remaining faces and interior represent the unconscious mind—deeper layers where non-linear integration takes place. The cube's continuous rotation symbolizes the mind's dynamic nature and the constant interaction between conscious and unconscious processes.

CHAPTER FIVE

The Invisible Wedge

A scientific narrative on perception, dreaming, lucidity, psychosis, and the contested borderlands of anomalous experience

Prologue: Where Experience Splits

There is a narrow interface in the human mind where **signal becomes meaning**. It is present in every waking moment, yet almost never seen directly. We notice the world, not the conversion—sound into audition, molecules into smell, photons into color, pressure into touch. But when that conversion is altered—by sleep, by trauma, by illness, by disciplined attention, or by experiences that appear to exceed ordinary explanation—the interface becomes visible. That interface is the invisible wedge.

The wedge is not a mystical object. It is a boundary condition: the place where perception is assembled, where memory is recruited, where emotion biases interpretation, where imagination fills gaps, and where the self-labels experiences as *real*, *imagined*, *dreamed*, *remembered*, or *hallucinated*. It is a seam between physiology and meaning, between the measurable and the lived.

This chapter develops a unified account of three states that expose the wedge most clearly: **waking perception**, **dreaming (including lucid dreaming)**, and **psychosis-spectrum phenomena** such as hallucination and delusion. It then extends outward into the cultural and scientific controversy surrounding **precognition**, **clairvoyance**, and **remote viewing**, and finally addresses why quantum experiments—especially the **double-slit experiment** and **entanglement**—so often become metaphors for consciousness.

A core tension guides the narrative. On one side stands the demand for scientific rigor: reproducibility, controls, alternative explanations, and humility in the face of uncertainty. On the other side stands the undeniable fact that some experiences arrive with an authority that feels absolute—especially when an event in a dream or vision appears to align with a later reality. Where does one place such experiences—particularly when the experiencer has been told that the only acceptable interpretation is *illness*?

The author of this book has lived inside that tension. He reports a lucid dream that carried emotional force and apparent foreknowledge, and a waking vision that unfolded exactly as anticipated. He was told these were symptoms of schizophrenia, and that he was delusional. Yet, from his perspective, the experiences predicted actual future events, and the mismatch between his lived certainty and the diagnoses offered to him became the impetus for sustained inquiry into anomalous cognition and metaphysical claims.

This chapter does not ask the reader to accept an extraordinary claim uncritically. It asks something else: to look carefully at the wedge—at how minds construct reality, how minds can misclassify internal events as external, and how institutions can misclassify extraordinary reports as pathology without adequate epistemic care. If the wedge is invisible in ordinary life, it becomes painfully visible when one's most intimate experiences are disputed by others.

I. The Architecture of Perception: From Stimulus to Experience

Perception begins with a deceptively simple fact: **the brain never touches the world directly**. It touches patterns of neural activity that *stand in* for the world. The taste of apple pie, the warmth of sunlight, the timbre of smooth jazz—these are not properties that travel from objects into consciousness. They are experiences assembled by a biological system from inputs that are physical, sparse, and coded.

1. Receptors: Specialized gates to the environment

External events—pressure oscillations, chemical compounds, electromagnetic radiation—interact with **sensory receptors**. These receptors are tuned. Auditory hair cells respond to mechanical vibration; olfactory receptors bind molecular features; photoreceptors respond to wavelengths of light. The receptors do not produce meaning. They produce **signals**.

2. Transduction: Turning energy into code

Through **signal transduction**, receptors convert physical stimulation into electrochemical changes that can be transmitted. This is an engineering marvel carried out in biological tissue: the conversion of one form of energy into another, with the goal of preserving information relevant to survival.

The nervous system then represents this information as patterns of action potentials—timed spikes and network dynamics. This conversion is neither neutral nor complete. It filters. It compresses. It emphasizes certain ranges and ignores others. A creature that evolved under specific constraints perceives those parts of reality that mattered to its ancestors.

3. Transmission and interpretation: The brain as an inference engine

Signals travel along neural pathways to specialized processing regions. The brain integrates incoming data with context, expectation, and memory. What emerges is not a raw readout but a **best-fit model**: the brain's working hypothesis of what is out there.

Perception, in this view, is a controlled hallucination constrained by sensory input—an inference that usually tracks reality because it has been refined by evolution and learning. But the same architecture that makes perception efficient creates its characteristic vulnerabilities: it can be fooled by illusions, biased by expectations, and in extreme cases can generate vivid perceptions without external causes.

The invisible wedge is already present here. It is the gap between stimulus and experience, between the external world and the internal construction. Most of the time, the construction is accurate enough that we mistake it for direct access.

II. Dreaming: When the World Goes Quiet but the Mind Does Not

Dreaming reveals the wedge by subtracting a key ingredient: **external constraint**. During sleep, particularly in phases when dreaming is most vivid, sensory organs still function in a basic physiological sense, but their integration into conscious experience is reduced. The environment becomes muted, and the brain's generative machinery takes the lead.

Dreaming is often described as a state where the sense organs are "turned off" while the brain remains active.

Technically, sleep involves altered patterns of thalamocortical gating, neuromodulation, and network connectivity. Phenomenologically, it feels like this: the theater lights dim on the outside world, and an interior world brightens.

1. The dream as synthesis

In dreams, the mind weaves together:

- **memories** (fragments of past events, faces, places, stories),
- **emotions** (fear, desire, grief, longing),
- **fantasies and simulations** (alternate outcomes, improbable scenarios),
- **residual stimuli** (sounds in the room, bodily sensations, temperature),
- and **self-modeling** (a continuing sense of "me," often simplified).

The brain does not simply replay memory. It constructs a scene that *feels* present. It populates that scene with narrative momentum. It assigns significance. It makes symbols.

2. Why dreams are so persuasive

Dreams can be bizarre, yet persuasive. They can violate physics, continuity, and identity, yet the dreamer rarely questions them. That is not because the dreamer is "irrational" in some moral sense. It is because the evaluative systems that, in waking life, constantly test reality are less dominant in ordinary dreaming.

The wedge widens: internal content gains autonomy. The brain is still doing what it always does—building a world-model—but now the inputs are predominantly internal.

III. The Mind's Eye: Training

the Generative System

The document's visualization exercise—closing the eyes, focusing on the space between the eyebrows, imagining a circle of light and allowing it to expand and transform—can be read as a deliberate engagement with endogenous imagery. Whether framed spiritually ("mind's eye") or cognitively (guided imagery), the practice reveals how attention can tune the brain's generative capacities.

1. A practical protocol for internal perception

The steps in the exercise function as psychological levers:

- **Posture and breathing** reduce noise and stabilize attention.
- **A simple seed image** (a circle of light) gives the mind something to hold without overdetermining content.
- **Gradual expansion** invites spontaneous emergence and pattern completion.
- **Detail sharpening** recruits perceptual resources—color, texture, motion—making the imagery more lifelike.
- **Projection outward** shifts imagery into a quasi-perceptual frame, as if viewing a screen.
- **Non-interpretation** prevents immediate story-making from smothering the phenomenon.

This matters for the chapter's larger argument: the mind can **generate vivid percepts** when properly cued. This is normal. It can be trained. It can feel real. And it can be experienced without external stimulus—yet it is not necessarily pathology.

IV. Lucid Dreaming: Awareness

Inside the Dream

Lucid dreaming is a special case because it partially restores what ordinary dreaming suspends: **metacognition**—the ability to know what one is experiencing and to reflect on it.

The document describes a classic induction technique: repeatedly asking, "Am I dreaming?" during waking life, so that the question arises within dreams. When it does, the dreamer trains a response that affirms lucidity: *Yes, I am dreaming. I am aware that I am aware.*

Lucidity is not simply waking up inside the dream. It is a hybrid state: the dream continues, but an additional layer of monitoring activates.

1. Two modes, one system

The document uses the language of a **subjective mind** and an **objective mind**. These terms can be understood functionally:

- The *subjective* mode: generative, associative, affect-driven, scene-building.
- The *objective* mode: reflective, evaluative, reality-testing, rule-checking.

In ordinary dreams, subjective generation dominates. In waking life, objective monitoring often dominates. In lucid dreams, the two modes overlap.

2. Stabilizing lucidity

Lucidity can be fragile. Excitement can wake the dreamer. Fear can derail control. The document's advice—remain calm, sustain attention, extend the experience—reflects a pragmatic truth: lucid dreaming is partly a skill of emotional regulation within an altered state.

Dream journals and recall training serve a similar function.

They increase the likelihood of noticing dream signs and support the formation of memory traces strong enough to persist into waking life. A dream not recalled might as well never have happened, yet dreams can influence mood and belief even when poorly remembered.

V. Dream Signs, External Cues, and the Threshold of Control

Lucidity can be triggered in multiple ways:

- **Internal questioning** (trained reality checks),
- **recognition of absurdity** (the mind notices the impossible),
- **external cues** (lights, sounds, tactile stimuli) that penetrate the dream.

The common denominator is interruption: a moment when the dream's narrative flow is briefly challenged by a signal that says, *reassess this state*. The reflective system gains purchase.

Here the wedge becomes an instrument. The dreamer learns to insert a tool into the seam between appearance and reality-labeling. In waking life, that seam is mostly sealed; in lucidity, it can be pried open.

VI. A Case Study in Haunting Lucidity: The Wave and the Birthday

At this point the chapter turns from general mechanisms to the kind of experience that drives people into the contested borderlands: a dream that does not remain merely a dream.

The author describes a lucid dream from youth:

He and his mother are on a beach. The weather is warm, the sky clear. A massive tidal wave appears on the horizon, moving toward shore. Panic seizes him; urgency floods the scene. He pleads with his mother—*we have to go, quickly!* —but she remains calm, unmoved, cleaning the windows of a small beach closet. She smiles and says, "Donnie, don't worry. It's just your father's birthday."

Two months later, his mother dies—on his father's birthday.

1. The phenomenology: why it imprints

This dream carries several features known to produce enduring psychological imprint:

- **high emotional arousal** (panic versus calm),
- **a clear, symbolic threat** (the wave),
- **a paradoxical reassurance** (it's just a birthday),
- **a later real-world anchor** (the date of death),
- **lucidity** (a sense of conscious awareness during the dream).

Lucidity matters because it adds a layer of *I was there, aware*—a quality that can make the memory feel more like an event than an image. The dream becomes a personal artifact, not merely a story the brain told itself.

2. Competing interpretations

A careful narrative must hold multiple hypotheses in view without forcing premature closure.

Psychological interpretations might emphasize:

- coincidence plus salience (birthdays are memorable markers),

- retrospective linking (after a death, the mind searches memory for foreshadowing),
- symbolic processing (waves as grief, birthdays as family binding points),
- memory reconsolidation (dream details can sharpen or shift after the fact).

Anomalous-cognition interpretations might emphasize:

- a genuine precognitive element,
- dream states as access points to information not ordinarily available,
- lucidity as a condition that increases "signal clarity."

The chapter's obligation is not to crown one interpretation as final, but to articulate what is at stake: the experience is meaningful and evidential *to the experiencer*, yet the standards of scientific proof demand more than narrative alignment.

Still, one cannot simply dismiss such a dream as irrelevant. The fact that it continues to haunt the author is itself data—data about how experiences shape belief, identity, and inquiry. The wedge is not only cognitive; it is existential.

VII. Waking Visions and the Social Cost of Interpretation

The author also reports a waking vision of future events that occurred exactly as predicted. Unlike a dream, a waking vision collides directly with ordinary reality-testing. If the prediction later matches events, the experiencer may treat it as confirmation of an extraordinary capacity; if others interpret it as pathology, the experiencer may feel erased.

1. The diagnostic dilemma

In clinical practice, hallucinations and unusual beliefs can be symptoms of schizophrenia-spectrum disorders, mood disorders with psychotic features, trauma-related conditions, substance-induced states, or other medical causes. Yet clinical assessment is not merely a checklist. It includes:

- the degree of distress,
- the impact on functioning,
- the flexibility of belief,
- the presence of other symptoms (disorganization, negative symptoms, cognitive decline),
- medical and substance history,
- cultural and spiritual context,
- and the individual's ability to reality-test in everyday life.

An experience being unusual is not sufficient for diagnosis. Nor is the claim, *I saw something before it happened*. What matters clinically is whether the person's relationship to the experience becomes impairing, frightening, rigid, or unsafe.

2. Epistemic harm

When someone reports a vision and is told, without careful nuance, that they are delusional, a secondary injury can occur: **epistemic harm**—the harm of being treated as an unreliable narrator of one's own life.

This harm can intensify suffering and encourage withdrawal. It can also push a person toward communities that offer validation without rigor. Thus the wedge can become a social fault line: the experiencer torn between dismissal and credulity, between institutions that deny and subcultures that affirm.

A mature inquiry needs a third path: validation of the *impact* without automatic validation of the *interpretation*.

VIII. Schizophrenia: Hallucination, Delusion, and the Biology of Misattribution

The document describes schizophrenia as a chronic disorder involving disruptions in thought, perception, affect, and behavior—often emerging in late adolescence to early adulthood—and affecting roughly 1% of the population. It highlights common symptoms: hallucinations, paranoid ideation, delusions of persecution or grandeur, disorganized thinking, negative symptoms, and cognitive impairment.

1. Hallucination as perception without stimulus

Hallucinations are perceptions experienced as real without corresponding external stimuli. Auditory hallucinations are most common, but visual, olfactory, tactile, and gustatory hallucinations occur.

At the level of mechanism, hallucinations can be understood as failures of **source monitoring**: the brain generates content (inner speech, imagery, memory fragments) and misattributes it as external. Neurochemical dysregulation—dopamine and other neurotransmitter systems—may bias the assignment of salience, making internally generated events feel urgent, important, or externally caused.

2. Delusion as belief fixation under uncertainty

Delusions can be framed as belief systems that become rigid and resistant to counterevidence, often forming around attempts to explain anomalous experiences. If a person hears a voice and cannot classify it as internal, the mind searches for an explanation. Under stress, with altered salience, explanations can crystallize into persecutory or grandiose narratives.

3. Treatment and dignity

The document notes that schizophrenia has no cure but can be managed with antipsychotic medications, CBT and related therapies, social supports, skills training, and sometimes hospitalization.

For this chapter's purposes, one ethical point is central: **explaining** hallucination biologically must not become a reason to **dehumanize**. The aim of science here is relief, understanding, and support—not reduction of a person to symptoms.

IX. Why Understanding Psychosis Matters—Even for the Study of "Paranormal" Reports

Understanding schizophrenia matters not only medically and socially, but conceptually. The wedge between signal and meaning becomes visible in psychosis because the mechanisms that usually keep the boundary stable are strained or altered.

For those investigating anomalous experiences, psychosis research supplies critical tools:

- models of source monitoring and misattribution,
- understanding of salience and prediction error,
- the role of trauma and sleep disruption,
- the ways community narratives shape interpretation,
- and the difference between unusual experience and

disordered life.

This is not a cynical move to "explain away" everything anomalous. It is a necessary discipline: before positing extraordinary mechanisms, one must understand the ordinary mechanisms that can produce extraordinary feelings.

X. Clairvoyance and Cultural Context: Respect Without Abdication

The document distinguishes prophecy/clairvoyance from schizophrenia and urges sensitivity. Cultural contexts shape how unusual experiences are interpreted. In some communities, visions are integrated into religious life; in others, they are pathologized.

A careful approach recognizes two truths:

1. People can have experiences that are meaningful within their culture and not inherently pathological.
2. Claims of clairvoyance are not scientifically established simply because they are culturally respected.

Clinical humility and scientific skepticism are not enemies. A clinician can respect a person's framework while still attending to distress and safety. A scientist can acknowledge the power of narrative while insisting on evidence.

XI. Precognition: The Claim, the Controversy, and the Cognitive Alternatives

Precognition—perceiving future events without sensory information—is one of the most persistent human claims, especially through dreams. The document notes that evidence remains inconclusive and that many reports can be explained via coincidence, hindsight bias, selective recall, and unconscious pattern recognition.

These cognitive mechanisms are not trivial; they are strong:

- **Base-rate neglect**: people forget how many dreams did *not* match anything.
- **Retrofitting**: after an event, vague dream elements seem specific.
- **Selective memory**: "hits" are retained; "misses" fade.
- **Pattern completion**: the brain predicts likely futures from subtle cues.

And yet, the author's motivation emerges precisely because some experiences feel *too exact, too timed, too aligned* to dismiss comfortably. This tension—between powerful cognitive explanations and compelling personal certainty—fuels both research and debate.

A responsible chapter therefore treats precognition as a **live claim**: culturally widespread, psychologically understandable, scientifically unverified, and personally consequential.

XII. Remote Viewing: History, Methods, and the Replication Problem

Remote viewing entered public consciousness partly through Cold War-era programs such as Project Stargate. Structured protocols attempted to reduce bias: viewers worked "blind," recorded impressions, and later compared them to targets.

The critical scientific obstacle has been **replicability** and **methodological robustness**. Some studies report statistically significant effects; critics argue for leakage, poor controls, selective reporting, and subjective scoring. Even proponents often struggle to produce consistent results across labs.

Remote viewing thus functions in this chapter as a case study in the wedge's social life: how institutional interest, human suggestibility, and the hunger for certainty can combine to produce an enduring legend even when evidence remains contested.

XIII. The Double-Slit Experiment: Observation, Measurement, and Misunderstanding

The double-slit experiment demonstrates interference when quantum systems evolve without which-path information, and the disappearance of interference when which-path information is obtained via interaction with a measuring device. It reveals wave-particle duality and superposition in a way that defies classical intuition.

The key point, often misunderstood, is this: in physics, "observation" means **measurement interaction**, not necessarily a human mind staring at the apparatus. A detector can collapse interference by entangling the system with the environment.

1. The measurement problem and interpretations

The measurement problem remains philosophically rich, and interpretations vary:

- **Copenhagen**: collapse upon measurement, without requiring consciousness.

- **Many-Worlds**: no collapse; outcomes branch.
- **Objective collapse**: collapse is a physical process.
- **Von Neumann/Wigner** style proposals: consciousness plays a role (controversial, minority).

The chapter should resist the temptation to turn physics into a proof of metaphysics. Quantum mechanics is strange enough without being used as a blank check for any claim about mind.

XIV. Consciousness and Quantum Mechanics: Metaphor vs. Mechanism

The document notes debates about whether consciousness affects quantum outcomes and references delayed-choice and quantum eraser experiments. These are often invoked to imply that the mind "creates reality." Most physicists do not adopt that conclusion.

Here the wedge appears again—this time as a conceptual wedge. People experience the mind as primary ("I know the world through consciousness") and thus find it intuitive that consciousness might be fundamental in physics. But intuitive metaphysics is not evidence.

Still, quantum theory can serve as a disciplined metaphor: it teaches that measurement context matters and that classical assumptions about definite properties can fail. The metaphor becomes dangerous only when it is mistaken for mechanism.

XV. Entanglement: Connection Without Communication

Quantum entanglement produces correlations between particles that cannot be explained by local hidden variables, as shown by Bell-type experiments. Measuring one particle constrains the predicted outcomes of measurements on the other, regardless of distance.

Entanglement feels like instant influence, but it does not allow faster-than-light messaging. It is a physical phenomenon with real applications—quantum cryptography, computing—yet it is often misappropriated as a catch-all explanation for psychic connection.

The careful position is this: entanglement is real, astonishing, and experimentally confirmed. Claims that it explains mind-to-mind information transfer remain speculative and unsupported.

Conclusion: Living at the Wedge Without Falling In

The invisible wedge is the boundary where **experience is generated and labeled**. In waking life it is stable enough to disappear. In dreaming it opens wide, letting the mind create worlds. In lucid dreaming it becomes navigable, allowing awareness to enter the dream's architecture. In schizophrenia-spectrum disorders it can become unstable in painful ways, allowing perceptions and beliefs to detach from external constraint.

And in anomalous experiences—especially those that seem to anticipate the future—the wedge becomes a battleground between two human needs: the need to explain and the need to be believed.

The author's lucid dream of the tidal wave and the birthday, followed by the death of his mother on that date, is not merely an anecdote. It is the emotional engine of inquiry. It demonstrates how a single experience can reorder a life: how it can make purely dismissive explanations feel insufficient, and purely credulous explanations feel dangerously easy. The later waking vision, experienced as exact prediction, deepens the dilemma and intensifies the question: *What is the mind capable of, and how would we know?*

A scientifically responsible chapter does not resolve this by proclamation. It resolves it by framing the landscape:

- by clarifying what perception is and how it can be fooled,
- by showing how dreams are generated and why lucidity matters,
- by distinguishing pathology from culturally meaningful experience without erasing either,
- by explaining why extraordinary claims struggle under rigorous testing,
- and by refusing to misuse quantum mechanics as a rhetorical weapon.

The wedge remains, and perhaps it must. It is the price of being a creature whose reality is mediated by a brain. But the wedge can be illuminated. And in that illumination, one may find something rarer than certainty: a method of inquiry that honors both the rigor of science and the gravity of lived experience.

CHAPTER SIX

The Touch of Reality

Why "Reality" Is Not Just What We See

In everyday language, reality sounds solid. It is the desk beneath our hands, the geometry of a room, the sound of a familiar voice, the scent of a hallway after rain.

Memory, Emotion, and the Classroom Climate That Makes Learning Possible

In our waking state we move through the world as though it is a stable, three-dimensional stage and our senses are faithful reporters. Sight, sound, touch, taste, and smell provide continuous input, and the mind—without our conscious effort—turns that stream of information into something we experience as coherent and true.

Yet the touch of reality is more than external facts. It is also the internal *felt sense* of what is safe, what is important, and what belongs to us. It includes the quiet meanings we assign to a teacher's tone, the micro-expressions on a peer's face, the way a classroom "feels" before anyone speaks. These subtleties may never rise to conscious awareness, but they are recorded, nonetheless. They become memory. And memory is never purely informational; it is braided together with emotion.

This is why the emotional climate of a classroom is not peripheral to learning—it is foundational. Understanding the relationship between memory and emotion is vital for creating effective educational environments. When teachers cultivate a nurturing, nonthreatening atmosphere, they reduce stress, enhance higher-level thinking, and encourage open communication. Students do not only learn *content* in such an environment; they learn that learning itself is safe, meaningful, and worth the risk of engagement.

This chapter explores how reality is "touched" and constructed in the mind—through conscious perception, unconscious processing, emotionally charged associations, and the social energy exchanged between teacher and student. It also examines why the same mechanisms used to influence consumers through advertising can be used ethically in education to support attention, meaning, and retention. The throughline is simple but profound: **what students remember depends greatly on what they feel while learning.**

I. The Waking World and the Unseen Recording Device

In waking life we experience ourselves as alert observers. We feel that we *choose* what to notice. But cognition is more layered than that. While the conscious mind experiences a curated version of reality, the unconscious mind processes far more than we can report.

A student may look at a whiteboard and remember a formula. At the same time, they may unconsciously register the teacher's patience, the pace of the lesson, the reactions of classmates, the social stakes of being called on, and the subtle cues that suggest whether mistakes will be punished or welcomed. The student's conscious recollection might be "We covered quadratic equations." The student's deeper memory might be "I felt tense," or "I felt seen," or "I learned I can ask questions without being ridiculed."

This distinction matters because education is not only about transmitting knowledge; it is about shaping the *conditions* under which knowledge can be encoded and retrieved. The brain does not store experiences as neutral files. Experiences are stored as patterns—sensory details, interpretations, and emotions bound together. When the emotional system flags an event as important—whether because it is joyful, frightening, humiliating, or inspiring—the memory system tends to treat that event differently. The moment becomes more "sticky," more retrievable, and more likely to influence later attention and behavior.

Subtle perception: what we don't "notice" still enters us

Consider the ordinary example of a billboard. A person may drive past it without deliberate attention and later find themselves drawn toward the advertised product. The conscious mind claims, "I wasn't influenced." The unconscious mind quietly replies, "You were exposed, you formed an association, and the association is now guiding preference."

Classrooms have their own billboards. They are not signs on the highway; they are daily cues that teach students what school means:

- the teacher's facial expression when a student struggles
- the class's laughter when someone answers incorrectly
- the teacher's patience—or impatience—when repeating instructions
- the unspoken rules about who belongs and who does not

- the rhythm of the day that either calms the nervous system or keeps it braced

These cues shape what students anticipate, and anticipation shapes attention. Attention shapes memory. Memory then shapes future anticipation. A loop forms, and once it forms, it can be difficult to break without intentional intervention.

Unconscious memory is emotional memory

The unconscious mind does not simply store impressions; it stores *felt meanings*. A student might not be able to articulate why they dread a subject, but their body knows. Another student might not be able to explain why they trust a teacher, but their nervous system relaxes when that teacher speaks. These are not mystical effects; they are the predictable results of learning systems designed to prioritize safety, belonging, and competence.

In this sense, the "touch of reality" in education is not only the material students handle or the words they hear. It is the emotional signature of the learning environment. Students are always asking, consciously or unconsciously:

- Am I safe here?
- Will I be shamed?
- Do I matter?
- Will my efforts count?
- Is curiosity allowed?
- Can I repair mistakes?

When the environment answers these questions with warmth and stability, stress decreases and cognition expands. When the environment answers with threat and unpredictability, stress rises and cognition contracts.

II. Dreams, Random Thoughts, and the Hidden Logic of Association

If waking perception shows how the mind builds a functional model of the world, dreams show how the mind builds worlds from within. During sleep, stored memories and emotions can appear as dreams that seem random or bizarre. A childhood home may appear alongside people from the present. A fear from years ago may borrow the face of a current friend. Dreams, though strange, reveal a key truth: **the mind organizes experience by association, and emotion is one of its strongest adhesives.**

This dreamlike mixing does not only happen at night. It also happens in daytime cognition whenever an unexpected memory surfaces. A smell, a song, a phrase, or a tone of voice can suddenly summon an old scene—sometimes vivid, sometimes only a faint mood. The conscious mind experiences this as a "random thought," but it is rarely random. It is the return of an association: a memory triggered by a sensory or emotional cue.

In classrooms, these associations matter because students come in carrying invisible histories. A teacher's raised voice—meant to restore order—may activate a student's earlier experiences of conflict. A quick correction—meant to be helpful—may trigger shame in a student who has been repeatedly embarrassed. Conversely, a calm presence and predictable structure can soothe students who live in chaotic environments outside school.

Teachers do not need to know every student's story to respond effectively. They need to understand a core principle: **the learner's internal world is present in the room, and it shapes what the learner can do with the lesson.** When teachers build emotional safety, they give students a better chance to separate the present from the past—to stay in the learning moment rather than being pulled into an old pattern.

The partnership of conscious and unconscious minds

We might imagine the conscious and unconscious minds as partners. The conscious mind scans, plans, chooses words, and solves problems. The unconscious mind absorbs cues, stores emotional impressions, recognizes patterns, and generates intuitions. In learning, these systems collaborate continuously.

When the collaboration is healthy, the unconscious supports the conscious. The student feels grounded and able to focus. When the collaboration is strained—often through chronic stress—the unconscious floods the conscious with threat signals: worry, vigilance, or avoidance. The student may still *want* to learn, but their cognitive resources are diverted toward emotional management.

A nurturing classroom does not eliminate all stress; learning requires challenge. But it distinguishes **productive challenge** from **threat**. Productive challenge signals, "This is hard, but you can do it, and you will be supported." Threat signals, "This is hard, and if you fail, you will be diminished." The first invites persistence and curiosity; the second invites shutdown, aggression, or escape.

III. Lessons from Advertising: Emotion, Memory, and Ethical Influence

Television commercials are built on a simple understanding: what we feel determines what we remember, and what we remember determines what we choose. Advertisers frequently aim to create strong emotional connections because emotion makes messages memorable and persuasive.

Commercials that evoke happiness, nostalgia, excitement, or sadness often remain in the mind longer than commercials that only list product features. Storytelling draws viewers in. Relatable characters create identification. Music and visual imagery intensify mood. Repetition strengthens association. Psychological principles like priming and conditioning ensure that the product becomes linked with a particular feeling.

This is not merely a marketing trick; it is a demonstration of how memory works.

What education can learn—without becoming manipulation

Education, too, creates associations. The difference is ethical intention. Advertising often seeks to influence behavior for consumption. Education seeks to influence behavior toward understanding, growth, and capacity. But the same psychological mechanisms are involved:

- **Story** gives information meaning.
- **Relatability** builds engagement and attention.
- **Emotion** strengthens encoding and retrieval.
- **Music, imagery, and novelty** can anchor memory.
- **Priming** can prepare students to interpret a lesson constructively.

A teacher who begins class with a calm routine—welcoming students by name, offering a brief settling moment, or previewing the day's goals—is "priming" students for clarity and safety. A teacher who frames an assignment as an opportunity to explore rather than a test to survive is shaping emotional expectation. A teacher who uses a compelling narrative in history or science is doing what the best commercials do, but in service of understanding rather than consumption.

The guiding ethical question is not whether emotion will be involved—emotion is *always* involved. The guiding question is whether emotion is being used to **support agency, dignity, and learning**, or to bypass judgment and pressure compliance.

Emotional connection is not entertainment—it is access

Some educators worry that engaging emotionally means "watering down" content. But emotional connection does not replace rigor; it enables it. When students care, attention increases. When attention increases, working memory has more usable capacity. When students are not bracing against threat, they can reflect, synthesize, and evaluate—skills at the heart of higher-level thinking.

A nonthreatening classroom does something critical: it converts emotional energy from defense into inquiry.

IV. The Brain's Learning Partnership: Emotion and Memory at Work

The relationship between memory and emotion is not a poetic idea; it is a functional reality of how the brain learns. Emotional processes and memory processes are intertwined. When emotion is activated, memory consolidation and retrieval can change dramatically.

Students often remember lessons that were engaging, meaningful, and relational. They may forget lessons delivered in a flat or stressful context, even if the content was important. This is not laziness. It is the brain doing what brains do: prioritizing what seems significant.

Stress narrows cognition; safety broadens it

A threatening environment—whether created by harsh criticism, unpredictable discipline, social humiliation, or chronic pressure—raises stress. Stress can interfere with concentration, working memory, and flexible thinking. Students may become hypervigilant, distracted, or avoidant. They may appear "unmotivated" when they are actually overloaded.

A nurturing environment reduces unnecessary stress, creating conditions where:

- attention can stabilize
- mistakes can be processed as feedback rather than danger
- students can ask questions without fear
- discussion becomes possible
- creativity and analysis can coexist

This is why the theme of this chapter is not optional: **when teachers cultivate a nurturing, nonthreatening atmosphere, they reduce stress, enhance higher-level thinking, and encourage open communication.** These outcomes are not soft ideals; they are prerequisites for deep learning.

Emotional intelligence belongs in academic learning

If emotions shape memory, then helping students understand emotions is not a side project—it is part of the learning infrastructure. When students can name what they feel, recognize stress cues, and use coping strategies, they gain control over attention and behavior. This supports academic achievement as well as well-being.

Simple practices can support this integration:

- brief reflection prompts ("What was challenging? What

helped you persist?")

- teaching language for emotions and needs
- normalizing nervousness before presentations
- showing students strategies for recovery after mistakes
- incorporating short mindfulness or breathing exercises to reset attention

None of these practices reduce academic expectations. Instead, they increase the likelihood that students can meet them.

V. Engagement as Relationship: The Teacher–Student Connection

Learning is not only an individual cognitive event; it is a social one. The "connection" between teacher and student acts like a bridge over which attention, meaning, and trust can travel. Students who feel engaged and connected to an instructor tend to absorb information more efficiently. Students who feel disconnected often struggle—not always because the material is beyond them, but because the emotional conditions needed for sustained attention are missing.

Rapport and trust are cognitive resources

Rapport is sometimes treated as a personality trait. In reality, rapport is a learning tool. When students believe that a teacher cares about their success and well-being, they are more likely to persist, ask questions, and accept feedback. Trust makes correction tolerable. It turns feedback into guidance rather than threat.

Teachers build trust through consistent, concrete actions:

- greeting students warmly

- acknowledging effort, not just performance
- giving constructive feedback with respect
- listening attentively
- following up on concerns
- keeping expectations clear and fair

These actions signal safety. Safety signals allow the brain to allocate resources to learning instead of defense.

The cost of distraction and the roots of disengagement

Distraction is not only a failure of willpower. It can come from personal stress, unclear instruction, social tension, or uninteresting content. But even "uninteresting content" is not always about entertainment; it is often about *meaning*. Students focus when they can locate purpose: how a lesson connects to their lives, their goals, or their curiosity.

A teacher can help by making relevance explicit, by designing interactive lessons, and by allowing students to participate rather than only receive. Discussions, group work, hands-on activities, multimedia, and structured peer collaboration can all promote engagement. Importantly, these methods are most effective in a nonthreatening environment where students feel safe enough to speak, risk an answer, and revise their thinking.

Active listening as a classroom technology

Active listening—genuine attention to students' ideas and concerns—does more than improve mood. It creates feedback channels. When teachers listen carefully, they gain real-time information about comprehension and misconception. They can adjust instruction on the spot. Students also learn that their minds matter, which strengthens motivation and participation.

In a classroom built on open communication, students are more likely to say:

- "I don't get it yet."
- "Can you explain that another way?"
- "Here's what I think—am I close?"
- "I disagree, and here's why."

These sentences are signs of intellectual life. They appear more readily when the environment is emotionally safe.

VI. The Eagle and the Nestlings: A Parable of Attention—and Its Limits

The document offers an analogy: an eagle feeds the nestling that seems most eager, and the eaglet that chirps loudest and stretches furthest is fed first. Applied to the classroom, the analogy suggests that engaged students naturally draw more teacher attention, forming stronger connections and receiving more support.

There is truth here. Teachers, being human, respond to visible cues of engagement: raised hands, eye contact, nodding, thoughtful questions, diligent notetaking. These cues signal readiness, and readiness invites investment.

But the analogy also invites a necessary refinement. In education, we cannot allow eagerness alone to determine nourishment. Some students are quiet not because they are disengaged, but because they are anxious, culturally conditioned to be reserved, processing in a second language, or recovering from repeated experiences of shame. Some students "stretch their neck" less because the environment has taught them that reaching is dangerous.

A nurturing classroom therefore uses the analogy carefully: it recognizes the power of visible engagement **while also building**

structures that feed the quieter learners. The goal is not to reward the loudest chirp; it is to cultivate conditions where more students can safely chirp at all.

What engagement looks like—and how to invite it

Engagement can be expressed in many forms:

- eye contact, nodding, leaning forward
- taking notes that capture key ideas and questions
- asking clarifying questions
- contributing to small-group discussion
- completing quick formative checks
- explaining an idea to a peer

Teachers can invite engagement by lowering the social risk. Instead of asking questions only to the full class, a teacher might use pair-share, anonymous response methods, or small-group roles. Instead of treating silence as defiance, a teacher might treat it as information and offer a different entry point.

When students begin to experience success—when their participation is met with respect rather than judgment—their emotional memory of school changes. And when emotional memory changes, engagement often follows.

Knowledge, memory, and emotion: the deeper meaning of the parable

The document rightly emphasizes that knowledge is tied to memories and that memories are intertwined with emotions. A moment of encouragement can anchor a concept in a student's mind. A moment of shared laughter can turn a lesson into something retrievable. The teacher's voice, enthusiasm, and manner are not superficial; they become part of what is stored.

In that sense, the eagle parable points toward a practical instructional truth: **students often learn best when they feel personally addressed by the learning environment.** That "address" can be direct attention, but it can also be structured inclusion: routines that ensure everyone is seen, strategies that distribute participation, and feedback that honors effort.

VII. Notetaking, Memory, and the Metaphor of "Psychic Connection"

The document uses a vivid image: students as "psychic vampires" feeding on the knowledge of instructors through notetaking. While the metaphor is intentionally dramatic, the underlying idea can be refined into a grounded educational principle: **note-taking is a form of active attention that strengthens memory and allows students to reconnect with the teacher's thinking later.**

A teacher's explanations are not merely information; they carry the teacher's organization of the subject—years of experience distilled into sequence, emphasis, analogy, and example. When students take notes, they capture not only content but also structure: what the teacher highlighted, how ideas relate, where confusion tends to occur, and what language makes the concept graspable.

Notes as a bridge back to the learning moment

When students review notes, they often recall more than words. They recall the tone of the class, the moment of insight, the example that made something click. This is the emotional component of memory at work. Notes become cues that reactivate the learning context—especially when the classroom environment was supportive and engaging.

In a threatening environment, notes may become a record of survival: "memorize this to avoid failure." In a nurturing environment, notes become a record of meaning: "I understand why this matters, and I can return to it."

The cognitive benefits of notetaking

Refined and stated plainly, the benefits include:

- **Sustained attention:** writing anchors focus and reduces drift.
- **Encoding through processing:** transforming speech into notes requires selection and organization.
- **Personalization:** students can add reflections, questions, and connections.
- **Retrieval support:** notes provide cues for later recall and application.
- **Structure:** notes create an organized map of a lesson, reducing study anxiety.

Teachers can strengthen these benefits by teaching notetaking explicitly—showing students how to capture main ideas, examples, and questions, rather than transcribing everything. They can also provide partial outlines or "skeleton notes" to reduce cognitive overload, especially for younger learners or students with learning differences.

The metaphor of psychic connection becomes, in refined form, a pedagogical truth: **students learn when they can enter the teacher's way of organizing knowledge—and then make it their own.**

VIII. The Classroom Feedback Loop:

Energy, Meaning, and Social Progress

The document describes a symbiotic relationship: engaged students generate positive energy that motivates teachers; motivated teachers communicate more effectively; learning improves; society benefits through educated individuals who innovate and lead. This is not a sentimental claim. It is an accurate description of how human systems work. Energy, attention, and meaning are contagious.

Engagement is mutual

When students participate—through questions, facial expressions, and body language—they provide feedback. Teachers adjust in response. Confusion becomes visible and addressable. Enthusiasm becomes fuel. The classroom becomes a living system rather than a one-way broadcast.

In a nonthreatening environment, this feedback is richer because students are less afraid to reveal misunderstanding. A teacher can correct misconceptions early, and students can refine thinking without shame. Over time, the classroom develops a shared culture: we can wrestle with ideas here; we can disagree respectfully; we can revise our positions; we can be wrong and still belong.

Education as awakening: expanding beyond the visible

The chapter's title—*The Touch of Reality*—invites a broader view of education. Reality is not only what is tangible. It includes the world of ideas, emotions, and memories that define human experience. Literature, art, philosophy, and history all invite students to step beyond immediate perception into the shared interior world of humanity.

When education is reduced to memorization under pressure, students may meet requirements without expanding awareness. But when education includes critical thinking, creativity, and emotional intelligence, it can awaken curiosity and deepen self-understanding. A thought-provoking discussion can open new perspectives. A respectful debate can teach intellectual humility. A well-guided reflection can help students understand their own emotional patterns—patterns that shape attention and learning.

In this sense, education touches reality in two directions at once: it helps students understand the external world and it helps them understand the internal world that interprets the external one.

The moral dimension: dignity as a learning condition

If memory and emotion are intertwined, then dignity is not merely an ethical preference; it is an instructional necessity. Students whose dignity is preserved can take cognitive risks. Students who are regularly shamed or dismissed may protect themselves by disengaging, acting out, or refusing to try.

A nurturing classroom is not indulgent. It is principled. It communicates:

- You are safe enough to think.
- You are respected enough to speak.
- You are capable enough to improve.
- You belong even when you struggle.

These messages do not lower standards; they build the human foundation that standards require.

IX. Practical Synthesis: Building the Nonthreatening Classroom Where Memory Can Thrive

The theme of this chapter can now be expressed as a practical blueprint. If teachers want students to remember more, think more deeply, and participate more openly, they must shape the emotional conditions of learning as intentionally as they shape the curriculum.

1) Reduce avoidable stress; keep productive challenge

- Make expectations clear and consistent.
- Use routines that stabilize attention at the start of class.
- Correct with respect; separate the learner from the error.
- Avoid public shaming; use private redirection when possible.
- Offer multiple chances for practice before high-stakes evaluation.

2) Make meaning visible

- Explain why a lesson matters and how it connects to larger goals.
- Use stories, examples, and analogies to create emotional anchors.
- Invite students to connect material to their experiences—carefully and voluntarily.

3) Encourage open communication

- Treat questions as evidence of thinking, not interruptions.
- Use structures that distribute participation (pair-share, small groups, written responses).
- Model intellectual humility: "Let's check that," or "Good point—let's revise."

4) Teach learning behaviors explicitly

- Teach note-taking strategies and provide scaffolds.
- Teach students how to study through retrieval practice, spacing, and reflection.
- Teach emotional regulation basics: breathing, self-talk, and recovery after mistakes.

5) Protect the relationship as part of the curriculum

- Learn students' names and pronounce them correctly.
- Acknowledge effort and growth.
- Follow through consistently; trust is built by reliability.

When these practices become part of classroom life, the "touch of reality" changes for students. Reality no longer feels like a test of worth. It feels like a place where thinking is possible. That shift—subtle but profound—changes the memory-emotion loop in the learner's brain. It makes knowledge more retrievable, discussion more alive, and higher-level thinking more attainable.

Closing: The Touch That Teaches

Students live in two worlds at once. One is the visible, three-dimensional world of desks, books, screens, and voices. The other is the invisible world of memory and emotion—associations formed from every success, every embarrassment, every moment of being understood or overlooked. Education happens at the intersection of these worlds. It happens where perception becomes meaning and where meaning becomes memory.

The classroom, then, is not merely a place where information is delivered. It is a place where reality is *touched*—not only by the hands, but by the nervous system. When teachers cultivate a nurturing, nonthreatening atmosphere, they reduce stress, enhance higher-level thinking, and encourage open communication. Students become more willing to engage, more able to concentrate, and more likely to remember—not just the facts, but the confidence that they can learn.

That confidence may be the most durable lesson of all.

CHAPTER SEVEN

Planetary Self-Awareness and the Missing God - Vulcan

To explore the human condition with any honesty is to admit that we are layered beings. We are instinct and reason, appetite and restraint, impulse and conscience.

Prologue: A Map Within the Mind, a Mirror in the World

We are also biology—cells, nerves, chemistry, and the electrical orchestration of the brain. Yet we speak about ourselves as though we possess a single, unified "I," when daily life reveals a more complex parliament of competing voices: the hunger that insists, the fear that warns, the mind that plans, the conscience that judges, the spirit that longs for meaning.

One of the most enduring attempts to describe this inner complexity is the classical triad of **id**, **ego**, and **super-ego**. Another powerful lens is anatomical: the **brain stem**, the **midbrain**, and the **cerebrum**, each representing distinct layers of function, from basic survival regulation to higher cognition and abstract reasoning. When these two models are placed side by side, a compelling parallel emerges: psychological "parts" appear to echo biological "levels," as though the psyche is not floating above matter but is shaped by matter—organized by the brain's architecture, refined by experience, and stabilized by social life.

But this chapter's ambition extends beyond the individual. If the human psyche can be described as a layered system—and if the human brain is also a layered system—then perhaps complex human **societies** can be understood in similar terms. Nations, cultures, and global alliances behave as though they, too, have instinctual drives, practical mediating structures, and idealistic or moral aspirations. At the scale of civilizations, we witness survival needs, competition, and territoriality; we also witness diplomacy, commerce, and coordination; and we witness ideals —law, human rights, scientific inquiry, and visions of what humanity *ought* to become.

When we extend the pattern further, the speculation becomes grand: if individuals and societies can be seen as layered intelligences, might the planet itself—through the interconnection of its peoples—be approaching a form of **planetary self-awareness**? And if so, what symbolic image best captures such a transformation?

The document that shapes this chapter offers a striking answer: **Vulcan**—Hephaestus to the Greeks—the Roman god of fire and metalworking, the divine artisan who transforms raw material into tools, weapons, and wonders. Vulcan is not merely a mythological character here; he becomes a metaphor for an emerging planetary mind: forged through adversity, disciplined by constraint, awakened by ingenuity, and capable—at last—of building something new from the elemental forces of existence.

I. The Psyche and the Brain: A Parallel Architecture of Being

The first parallel is intimate: the relationship between the **human psyche** and the **physical brain**. Classical psychoanalytic language describes the mind in three interacting systems.

The Id: the primal engine

The **id** is the most primal and unconscious part of the psyche. It operates on the **pleasure principle**, seeking immediate gratification of basic drives: food, comfort, warmth, affection, sex, relief from pain, and the reduction of tension. It does not argue politely; it demands. It does not negotiate with tomorrow; it insists on now.

In daily life, the id is not "evil." It is foundational. Without it, there is no hunger that leads to nourishment, no desire that leads to reproduction, no instinct that leads to protection. But without constraint and direction, the id can become chaotic—overconsuming, impulsive, or destructive.

The Ego: the mediator of reality

The **ego** develops as the psyche meets the world. It operates on the **reality principle**, mediating between instinctual demands and external constraints. The ego negotiates, delays gratification, makes plans, tests reality, and chooses strategies. Where the id says, "I want," the ego asks, "What is possible, safe, and wise here?"

The ego is not merely rational; it is adaptive. It can be creative, strategic, and resilient. It is the part of us that learns rules, reads situations, and builds workable lives.

The Super-ego: the internal judge and ideal

The **super-ego** represents internalized moral standards and societal ideals. It judges behavior as right or wrong, worthy or shameful, admirable or unacceptable. From the super-ego come feelings of **guilt** and **pride**, the desire to align with values, and the longing for perfection.

The super-ego can elevate. It can also persecute. A healthy super-ego guides; a harsh one punishes. But either way, it pushes the psyche beyond mere survival toward meaning, belonging, and moral identity.

The physical brain: stem, midbrain, cerebrum

Now set this psychological triad alongside a simplified anatomical triad:

- The **brain stem** governs essential autonomic functions—heart rate, breathing, arousal—and supports reflexes and primitive survival instincts.

- The **midbrain** helps coordinate motor movement, processes auditory and visual information, and regulates arousal and alertness, acting as a bridge between primitive functions and higher processing.

- The **cerebrum** enables higher cognition: thought, memory, emotion, decision-making, planning, abstract reasoning, and—crucially—complex social and moral consideration.

The alignment: id / brain stem; ego / midbrain; super-ego / cerebrum

The chapter's central proposition is not that Freud "mapped" perfectly onto neuroscience, but that the **structure** of the analogy is illuminating.

- The **brain stem**, like the **id**, is concerned with immediate survival needs and functions largely outside consciousness.

- The **midbrain**, like the **ego**, integrates information and guides action in response to reality.

- The **cerebrum**, like the **super-ego**, supports advanced reasoning, planning, and moral judgment—those forms of thought that allow us to live not only as organisms but as responsible persons.

Seen this way, the psyche is not a ghost in the machine; it is the lived experience of a machine that has become self-reflective. Biology does not merely enable psychology; it shapes its very form. And psychology, in turn, shapes biology through habit, stress, learning, and culture.

This is why the parallel matters: it offers a **unified** picture of the human being—mental and physical not as enemies, but as inseparable aspects of one integrated self.

II. An Archetype of Three Planes: Instinctual, Physical, Spiritual

From this parallel emerges a broader "ethereal archetype," described as existing on **three planes of existence**. Whether one reads these planes literally or symbolically, they provide a useful map of the human experience.

1) The Instinctual Plane

This is the domain of primal urges: hunger, thirst, safety-seeking, sexuality, territoriality, and attachment. In psychological terms, it corresponds to the **id**; in anatomical terms, to the **brain stem**. It is the plane of life-force and urgency.

In education, politics, and personal relationships, this plane appears whenever fear overrides reflection or whenever appetite overwhelms restraint. It is not a "lower" plane to be despised; it is a base to be understood and guided.

2) The Physical Plane

This plane concerns sensory processing, action, and engagement with reality. It corresponds to the **ego** and the **midbrain**—the mediators and integrators. The physical plane is where perception meets choice: how we move through space, interpret signals, respond to threats, and make practical decisions.

This is the plane of adaptation: learning a craft, solving problems, negotiating roles, managing conflict, building systems that work.

3) The Spiritual Plane

Here "spiritual" is not restricted to religion. It refers to higher-level meaning-making: moral reasoning, ideals, conscience, aspiration, and the search for purpose. This plane corresponds to the **super-ego** and the **cerebrum**, where abstract thought and ethical judgment can flourish.

This plane asks: *What should we become?* It seeks coherence between action and values. It builds civilizations, laws, art, and visions of justice.

A unified archetype

The archetype living on all three planes captures what we recognize in ourselves: we are at once animal, citizen, and seeker. We are driven, practical, and idealistic—often in conflict, sometimes in harmony. Understanding the interplay of these planes is a step toward personal integration.

But the chapter asks for more: can this model scale up?

III. Societies as Psyches: Superpowers, Emerging Nations, Primitive Cultures

If individuals contain layered systems, societies do as well. Civilizations appear to enact instinctual drives, practical mediations, and moral aspirations—sometimes in noble form, sometimes in distorted form. The document proposes a provocative mapping:

- **Primitive cultures** align with the **brain stem / id**
- **Emerging nations** align with the **midbrain / ego**
- **Superpowers** align with the **cerebrum / super-ego**

To treat this mapping responsibly, it must be framed carefully. "Primitive" cannot be equated with lesser humanity; every culture possesses intelligence, dignity, and complexity. The mapping is not about value—it is about *function and emphasis*.

Primitive cultures: the foundational instincts of a collective

Traditional cultures that maintain older ways of living often preserve essential human knowledge: survival skills, ecological awareness, kinship structures, rituals of meaning, and the continuity of identity through time. In the model, they represent a society's foundational layer—the "instinctual memory" of the species.

They remind the global system that life depends on food, water, land, seasonal rhythms, birth and death, and the bonds of community. They ground us.

Emerging nations: the mediators of development

Emerging societies, modernizing and balancing competing demands, resemble the ego-like function of mediation. They negotiate between immediate needs—security, infrastructure, employment—and higher aspirations—education, rights, innovation, and global participation.

They are the "bridge layer": practical, adaptive, sometimes unstable, often striving. They embody the tension of transformation.

Superpowers: the higher-order complexity of collective mind

Superpowers, in this analogy, represent advanced cognitive structures: large-scale organization, complex technologies, sophisticated institutions, strategic planning, and the shaping of global norms. They approximate the "cerebral" function of civilization: abstraction, long-range coordination, and the assertion of ideals.

Yet the analogy also exposes a danger. A superpower can claim moral authority (super-ego) while acting from primal fear (id). Nations can rationalize aggression with lofty language. They can confuse domination with virtue. So the model is descriptive, not celebratory. It shows what is *possible* at a high level of organization, but it does not guarantee wisdom.

The world as a layered intelligence

When these layers interact—traditional stability, developmental mediation, and advanced coordination—a global system begins to resemble a single organism. Conflicts arise when layers miscommunicate, exploit one another, or refuse mutual respect. Progress occurs when layers recognize interdependence.

From this point, the chapter moves toward its most visionary claim: **planetary self-awareness**.

IV. Planetary Self-Awareness: The Earth as a Mind in Formation

Imagine the planet not as a rock populated by separate groups, but as a living field of interaction capable of developing a unified consciousness. In this view, each societal "layer" contributes to the formation of a global mind:

- Superpowers contribute complex planning, ethics, and innovation—the "higher thought" of the planetary system.

- Emerging nations mediate between necessity and aspiration—turning ideas into living systems, bridging gaps, balancing growth.

- Traditional cultures preserve the foundational instincts and memory of the species—keeping the planetary organism grounded in life's essentials.

This is not a claim that the Earth is already self-aware in a biological sense. It is a symbolic and philosophical model that asks: *What would it mean for humanity to think as one body?* What would it mean to recognize that the boundaries we defend are internal boundaries within a larger organism?

The document suggests that when planetary self-awareness emerges, humanity may experience a profound shift—a collective epiphany—leading toward what it calls **god-consciousness**. In that state, identity expands: one is not only a citizen of a country but an "earthling," a cell within a planetary mind.

This is where the symbolism of Vulcan becomes crucial, because planetary self-awareness is not imagined as gentle abstraction alone. It is imagined as something **forged**—through fire, labor, constraint, creativity, and transformation.

V. The Cosmic Theater: Gods as Symbols, Planets as Archetypes

Before Vulcan steps fully into view, the text invites us to look outward—to the solar system—and notice how Greco-Roman mythology has shaped the names of celestial bodies. The planets become a "cosmic theater," each carrying symbolic traits:

- **Helios/Sol (the Sun)**: central life-giver, core energy
- **Mercury/Hermes**: speed, communication, messages
- **Venus/Aphrodite**: love, beauty, attraction
- **Earth/Gaia**: life, nurture, home

- **Mars/Ares**: war, conflict, aggression
- **Jupiter/Zeus**: authority, leadership, order
- **Saturn/Cronus**: time, discipline, structure
- **Uranus/Ouranos**: heavens, innovation
- **Neptune/Poseidon**: water, depth, mystery

These are not scientific descriptors; they are cultural symbols. But symbols matter because they shape imagination, and imagination shapes direction. A civilization that imagines its universe as meaningful tends to seek patterns, narratives, and purpose. Myth becomes a bridge between what we can measure and what we can feel.

And then there is the absence—the missing god.

VI. The Missing God: Why No Planet Is Named Vulcan

Vulcan (Hephaestus) is a major deity: god of fire, metalworking, craftsmanship, invention. Yet no planet bears his name. This gap has long provoked curiosity, and history added fuel to the mystery.

In the nineteenth century, the mathematician **Urbain Le Verrier** proposed that irregularities in Mercury's orbit might be explained by an undiscovered planet inside Mercury's path. The hypothetical body was called **Vulcan**. Searches followed. No planet was found. Eventually, Einstein's theory of general relativity explained Mercury's orbital precession without requiring an additional planet.

Scientifically, "Planet Vulcan" dissolved. Symbolically, Vulcan remained potent. The missing planet becomes an image of the unseen: the hidden forge, the invisible chain, the absent piece that nonetheless organizes the story.

The document proposes that the search for Vulcan represents a deeper human quest: the pursuit of hidden truths—those forces not immediately visible that nonetheless shape our world.

VII. Vulcan's Origin: Rejection, Fall, and the Hidden Forge

Vulcan's myth is not a simple hero tale. It begins with pain.

The rejected child

Vulcan, known as **Hephaestus** in Greek mythology, is the son of **Hera**, queen of the gods. In one account, his appearance is so displeasing to Hera that she casts him out of Olympus. The rejection is immediate and absolute: the child is expelled from the realm of beauty and power because he does not fit its aesthetic demands.

In another account, Vulcan intervenes in a quarrel between Hera and Jupiter (Zeus). Jupiter, enraged, throws him from Olympus. In this version, the fall is punishment for loyalty or protest—an act of power meant to enforce hierarchy.

Both versions carry the same emotional core: **exile**. Vulcan is separated from the center.

The long descent and the sea's sanctuary

The myth describes Vulcan falling for an entire day before crashing into the sea. Whether one reads this as literal poetry or symbolic psychology, the image is unmistakable: an extended descent from the heights of privilege into the depths of uncertainty.

He is rescued—by the **Nereids**, sea nymphs—who hide him in underwater caves. The caves become a place of concealment and gestation, a womb-like refuge where rejected potential can develop outside the gaze of judgment.

The island of Lemnos: craft as redemption

In the hidden sanctuary—often associated with the island of **Lemnos**—Vulcan heals and works. He takes raw material from the earth and transforms it through skill and persistence. This is the core of his character: he is a creator whose power is not inherited status but **earned mastery**.

Here the myth becomes an allegory of transformation. The rejected child becomes the artisan. The exile becomes the forge. The deformity becomes not a mark of shame but a reminder that power can be built from pain.

Vulcan's workshop is where fire becomes not destruction but discipline—heat applied with purpose.

VIII. Return to Olympus: Recognition of True Worth

Eventually the gods recognize what they have discarded. They invite Vulcan back to the heavens—not out of pure compassion, but because his gifts are indispensable. He creates magnificent weapons, shields, devices, and artifacts. The heavens, for all their authority, require the labor of invention.

The narrative arc matters: Vulcan's redemption is not achieved by pleading for acceptance. It is achieved by becoming so skilled that his exclusion becomes costly to the very system that rejected him.

This is not merely a story about revenge or triumph. It is a story about the civilizational role of technology and craft. Societies, like Olympus, depend on creators. They may romanticize beauty and power, but they cannot advance without those who know how to shape the world.

IX. Vulcan's Revenge: The Golden Throne and the Invisible Chains

The myth does not sanitize Vulcan. His pain does not evaporate when he becomes useful. In one of the most famous episodes, Vulcan crafts a magnificent **golden throne** and presents it to Hera as a gift—seemingly benevolent, radiant, irresistible.

When Hera sits, she is ensnared by **invisible chains** that only Vulcan can unlock. None of the other gods can free her. The scene is both comic and frightening: the queen of the gods, trapped by a device created by the son she rejected.

The symbolism is sharp. The throne—an emblem of authority—becomes a prison. Beauty becomes danger. Power becomes dependence. The one dismissed as flawed demonstrates mastery over the system that judged him.

The gods plead. Jupiter recognizes the severity and the leverage. Vulcan has the advantage not because he commands armies, but because he commands **technology**—the power to create mechanisms that reshape reality.

In modern terms, this is the story of an innovator who, once cast out, returns with tools that even kings cannot counter.

X. The Marriage Bargain: Vulcan and Venus

Vulcan's demand for Hera's release is astonishing: **Venus (Aphrodite)** as his wife.

Jupiter consents, and the arrangement is framed as strategy. Jupiter believes that marrying Venus to Vulcan will distract her from Mars (Ares) and thereby reduce conflict among the gods. In this reading, marriage is not romance; it is a political instrument. Venus does not protest. The deal is sealed.

This episode reveals several intertwined themes:

1. **Leverage and negotiation:** Vulcan uses his unique power to obtain what he desires.
2. **The manipulation of relationships:** Jupiter treats marriage as governance.
3. **The collision of beauty and craft:** Venus embodies attraction; Vulcan embodies invention.

The myth suggests that creation seeks union with beauty, and beauty—often volatile—can be constrained or redirected by structure. Yet the story also implies tension: love and war (Venus and Mars) are powerful forces, and no contract can fully tame them. Vulcan's marriage is therefore both triumph and complication.

XI. Vulcan as Archetype: Invention, Technology, and the Balance of Forces

In the document's analysis, Vulcan becomes more than a character. He becomes an **archetype** of creative power.

The benevolent maker of weapons

Vulcan is a weapon maker, yet often portrayed as peaceful and benevolent. The contradiction is deliberate. Technology is neutral in itself; its moral direction depends on intention. The same skill can build plows or swords, medical devices or instruments of surveillance. Vulcan embodies that ambivalence: the ability to create what is beautiful and useful—and also what is dangerous.

The redeemed outcast

His return to Olympus symbolizes the recognition of true worth after rejection. It is a lesson in how societies often treat inventors: dismissing them until they become necessary, then celebrating them while forgetting the cost of their exile.

The mediator between Venus and Mars

A particularly striking symbolic interpretation appears in the text: Vulcan, by marrying Venus, becomes a force that separates Venus from Mars. In planetary symbolism, one might imagine Vulcan's "place" as residing between **love** and **war**, between attraction and aggression, between the forces that bind and the forces that break.

This is not astronomy; it is philosophy. It suggests that invention —the disciplined transformation of raw power—might be what prevents beauty from being consumed by conflict. Technology, in its best form, can become a mediator: not merely amplifying power, but structuring it, channeling it, civilizing it.

Tapping the collective unconscious

Vulcan is also described as drawing from a collective reservoir —the shared deep mind of humanity—to create art and technology. This casts him as the archetype of the creator who brings forth what many dimly sense but cannot yet build.

In that sense, Vulcan resembles the civilizational impulse itself: the drive to take what is hidden, raw, or chaotic and render it usable, meaningful, and enduring.

XII. Planetary Self-Awareness Forged in Vulcan's Image

Now the chapter's themes converge.

If the **id/brain stem** is the survival substrate of humanity, and if the **ego/midbrain** is the mediator of practical life, and if the **super-ego/cerebrum** is the domain of ideals and ethics—then the global human system can be seen as moving toward a higher integration of these layers.

Planetary self-awareness, in this model, is not the elimination of instinct or the triumph of pure morality. It is the **integration** of layers into a coherent whole. And that integration requires a forging process: heat, pressure, discipline, design.

That is why Vulcan is such a fitting image.

- He is born into power yet cast out—like humanity, capable of greatness but still wounded by its own judgments.
- He falls into the depths—like civilization, repeatedly descending into war, collapse, and suffering.
- He builds in the cave—like societies building knowledge in obscurity until it becomes transformative.
- He returns not as a supplicant but as a master—like humanity potentially returning to itself with mature capacities.
- He binds the throne with invisible chains—like technology binding and restructuring the rules of power.
- He marries Venus through a bargain—like civilization trying to tether beauty and love to structure, so that they are not perpetually seized by war.

If planetary self-awareness is coming, it may arrive through a collective realization—an **epiphany**—that our beliefs, identities, and moral frameworks are not merely inherited truths but learned patterns stored in memory and reinforced by culture. A planetary epiphany would be the moment humanity recognizes itself as one organism with many organs, one mind with many functions.

Epiphany as a collective turning

Beliefs shape identity. They guide fears and aspirations. They create the limits of what seems possible. When knowledge accumulates, the mind can suddenly synthesize it into a profound realization. That is epiphany: not a slow persuasion, but a sudden reorganization.

The document suggests that humanity could experience such a moment collectively—an evolutionary leap into heightened awareness, even "god-consciousness." Whether one reads this spiritually or metaphorically, the implication is the same: the next stage of human development may require an expanded identity, where "earthling" becomes a primary category of self-understanding.

And if that stage requires forging—transforming raw drives into disciplined creativity—then Vulcan is not missing at all. He is not absent; he is emerging as an inner necessity.

Epilogue: The Forge and the Future

A planet becomes self-aware not by wishing, but by building the conditions for awareness: communication across differences, ethical frameworks that restrain violence, institutions that coordinate resources, and narratives that expand belonging. The forging of a planetary mind will require the reconciliation of instinct with conscience, power with responsibility, innovation with wisdom.

Vulcan's myth endures because it speaks to this work.

He teaches that rejection can become mastery, that exile can become invention, that pain can become discipline, and that creation is a form of power more enduring than force. He also warns that creation can be used for binding as well as liberation, for revenge as well as healing. The forge makes tools; it does not decide how they will be used.

If humanity is approaching a planetary epiphany, it will not be a mystical escape from history. It will be a crafted outcome —an intelligence built from raw material: our instincts, our institutions, our ideals, our stories. In that sense, the rise of planetary self-awareness would indeed arrive in Vulcan's image: not as a soft awakening, but as a hard-won transformation —fire-touched, hammer-shaped, and finally capable of making something beautiful that can endure.

CHAPTER EIGHT

Touchstone

The Planet That Wakes, Sleeps, and Dreams

A touchstone is a simple object with an ancient purpose: it tests what is real.

Prologue: The Touchstone of a Living World

A touchstone is a simple object with an ancient purpose: it tests what is real. Gold is drawn across it to reveal whether it is true or adulterated, genuine or false. In a similar way, the idea of **planetary self-awareness** functions as a philosophical touchstone. It tests the limits of how we think about life, consciousness, and belonging. It asks whether awareness is confined to individual skulls—or whether awareness can emerge wherever countless living parts exchange information, respond to signals, and form a coherent whole.

If the human mind arises from the synchronized activity of billions of neurons, then a planet rich with life invites a daring analogy: perhaps the planet's life forms serve as sensory receptors and interpretive nodes for a larger, distributed awareness. Perhaps forests and oceans, birds and insects, corals and fungi, human cities and migratory paths are not merely inhabitants of Earth, but **organs of perception** within a planetary system that—through living matter—can experience itself.

This chapter refines that proposition into a structured narrative. It does not demand belief as doctrine; rather, it explores a framework: **Earth as a conscious system** whose awareness emerges from the complex web of life. Within that framework we can ask deeper questions:

- If Earth can be self-aware, does that awareness fluctuate?

- Does a conscious planet have **phases of wakefulness and sleep**, shaped by seasons, cycles, and long ecological rhythms?

- If the planet "sleeps," does it also **dream**—and if so, could it manifest a kind of dream or astral body expressed through symbolic natural events, cultural myths, and long-term planetary shifts?

- Are ice ages, warming periods, floods, storms, migrations, extinctions, and evolutionary bursts merely mechanical phenomena—or might they also be understood as expressions within a dreaming planetary mind?

- Could there be something like **planetary lucid dreaming**, in which a species—especially one capable of reflective thought—becomes aware that it participates in the planet's dream and learns to influence it?

- And finally: does a planet have **memories**—patterns that persist, repeat, and shape the future the way memory shapes an individual?

To hold these questions well requires both imagination and discipline. We must respect science while admitting that science does not exhaust meaning. We must avoid turning metaphor into careless certainty, while also refusing the impoverishment of thought that insists only the measurable is real. "Touchstone" is the right title, because the chapter's aim is to test what kind of relationship we might form with Earth when we dare to think of it not as a stage, but as a living, sensing, remembering system.

I. Planetary Self-Awareness: Consciousness as an Emergent Ecology

Planetary self-awareness, as presented in the source text, begins with a straightforward claim: **a planet becomes aware of its existence through the sensory receptors of the living matter it supports.** That is, a planet's awareness is not a centralized "mind" located somewhere beneath the crust; it is an **emergent property**—a distributed phenomenon arising from the countless interactions of life.

1. The planet as a network of sensing

Every organism perceives. A bird maps air currents and magnetic fields; a plant tracks light, moisture, and chemical signals; a wolf reads scent and movement; coral responds to water temperature and acidity; microbes interpret nutrient gradients; and humans—through vision, language, and abstract thought—construct models of the world. None of these perceptions exist in isolation. Ecosystems are informational webs: one species alters the environment, another senses the alteration, and a cascade of responses follows.

In this view, life is not simply "on" the planet the way moss grows on stone. Life is a **sensory layer**. It is as though the planet grows eyes, ears, skin, and nerves—millions of different kinds, operating at different scales—so that it can register its own conditions.

2. A consciousness that is distributed, not localized

Human consciousness emerges from a distributed system: no single neuron contains your identity, yet the whole network produces experience. Planetary awareness, if it exists, would be similar: not a single command center, but a **distributed field of perception** arising from innumerable living nodes.

This matters because it changes how we interpret "planetary consciousness." It is not a mystical ghost hovering over Earth. It is closer to an ecological mind—an intelligence in relationships, in feedback loops, in the self-regulating dance of systems.

3. Feedback loops: the planet experiences and responds

A planet "experiencing" reality is a metaphor, but it points to something concrete: organisms perceive conditions and respond, and in responding they modify conditions. Birds shift migration routes; plants change flowering times; microbes alter soil chemistry; humans build cities and technologies; coral reefs bleach and recover or collapse; forests regrow or retreat. These responses are **feedback loops**—the same kind of circular causality that makes a nervous system adaptive.

In this framework, planetary self-awareness is inseparable from planetary self-regulation. Awareness is not merely "knowing"; it is **sensing-and-adjusting**.

II. Gaia Hypothesis: The Earth as a Self-Regulating Organism

The Gaia hypothesis is often invoked when discussing planetary consciousness because it offers a scientific-adjacent bridge between metaphor and mechanism. In its broadest form, Gaia proposes that Earth and its biological systems behave like a **vast self-regulating organism**, maintaining conditions conducive to life through feedback processes.

The hypothesis does not require that Earth is conscious in the human sense. But it opens conceptual space for viewing Earth as more than inert matter. It asks us to see the biosphere, atmosphere, oceans, and geology as interlocking systems that can stabilize temperature, regulate chemical composition, and sustain equilibrium—sometimes over astonishing spans of time.

1. Homeostasis and planetary "health"

In a body, homeostasis maintains stable internal conditions: temperature, pH, oxygen levels. Earth exhibits analogous patterns: the carbon cycle regulates greenhouse gases; oceans absorb and release heat; forests and plankton transform carbon and oxygen; soils store nutrients; ice caps reflect sunlight.

These patterns can be described as physical chemistry. But they can also be interpreted as something more: the planet behaving **as if** it has a preference for survivable conditions. Whether we call that preference "life's tendency to persist" or "Gaia's regulation," the effect is the same: Earth functions like a system that resists collapse, until pressures exceed thresholds.

2. A moral implication: stewardship within a living system

If Earth is self-regulating, humans are not external engineers managing a passive resource bank. We are participants in the regulatory system—capable of amplifying stability or disrupting it. This perspective often produces a sense of responsibility. To harm ecosystems is not merely to damage scenery; it is to injure organs in a larger body that sustains us.

The Gaia lens also reframes human identity. We are not simply individuals competing on a rock. We are cells in a biospheric organism—nodes in a planetary network that can either cooperate with the whole or behave like uncontrolled growth.

III. The Rhythms of Awareness: Wakefulness and Sleep at Planetary Scale

The source text proposes that planetary awareness may be **cyclical**, like the cycles of wakefulness and sleep in organisms. This analogy becomes compelling when we consider that Earth is already rhythmic: days and nights, seasons, ocean currents, migration cycles, flowering cycles, predator-prey cycles. Even geological and climatic processes have long rhythms.

1. Seasonal wakefulness and dormancy

In spring and summer, many regions erupt into visible vitality: growth accelerates, reproduction proliferates, sensory exchange intensifies. In winter, activity often quiets: growth slows, many organisms enter dormancy or migration, ecosystems shift into lower metabolic states.

If a planet's awareness is an emergent product of biological activity and interaction, then it is plausible—at least metaphorically—that awareness has **phases of intensity**. When life is most active, the planet "feels" more. When life is dormant, awareness "dims."

This does not mean the planet becomes unconscious in winter; rather, the signal-to-noise ratio changes. Inputs reduce. Interactions slow. A different mode of processing may prevail.

2. Deep cycles: glaciations and warming periods

Beyond seasons are planetary cycles measured in thousands to millions of years: ice ages and interglacials, shifts in atmospheric composition, reorganizations of ocean circulation, tectonic rearrangements that reshape climate and biodiversity. These deep cycles could be interpreted as **long sleep phases** and **long wake phases**, or as slower waves within a vast mind.

In human sleep, the brain is not inactive. It performs consolidation, repair, and integration. If we allow the analogy, Earth's long dormant periods might not be simple absence—they might be phases of reconfiguration.

3. Awareness as "state," not constant

Human consciousness varies: alertness fluctuates, attention narrows and widens, emotion colors perception, fatigue dulls processing. Earth's biosphere also fluctuates: resilience rises and falls, diversity expands and contracts, stability holds and breaks. The planetary model encourages us to see these fluctuations as shifts in **state**—not merely random events, but expressions of a living system transitioning through modes.

IV. The Planet's Dream: Does Earth Manifest an Astral Body?

If the planet can "sleep," the next question follows naturally: **does it dream?** The text proposes an even bolder image: a planet might manifest a **dream body**, an astral expression—symbolic, creative, and strange—through which it explores possibilities.

This is not a scientific claim; it is a mythopoetic framework designed to make sense of profound environmental changes and cultural intuitions. Yet even as metaphor, it invites rigorous thought.

1. What could a planetary dream be?

Dreaming in humans is not merely random imagery; it is a form of processing. It integrates memory, emotion, and learning. It can replay threats, rehearse social scenarios, test alternatives, and reorganize experience into narrative.

At planetary scale, a "dream" could mean periods when Earth's systems produce **novel configurations**—new climate regimes, new ecological arrangements, new evolutionary experiments. Dreams are the mind's sandbox; a planetary dream would be the biosphere's sandbox.

2. The dream body as symbolic natural phenomena

The text suggests that the planet's dream body might be "visible" through symbolic representations: unusual weather patterns, rare natural events, sudden ecosystem shifts, or large-scale geological activity. These events would not be "messages" in a literal sense, but expressions of a deep system transitioning—like symbolic imagery rising from the subconscious.

Under this framing:

- **Volcanic eruptions** could resemble emotional surges—release of pressure from below.
- **Earthquakes** could resemble abrupt neural reorganization—fault lines shifting.
- **Floods and storms** could resemble overwhelming affect—systems overflowing their channels.
- **Mass migrations** could resemble dream logic—sudden movements, instinct-driven relocations in response to unseen cues.

The danger here is romanticizing suffering. Floods and tornadoes devastate lives. The dream-body metaphor must not trivialize pain. Its role is not to justify catastrophe but to explore how a conscious-system model might interpret large-scale transitions without reducing Earth to dead machinery.

3. Cultural myth as a mirror of planetary dreaming

The text also proposes that human myths, shared symbols, and collective imaginations might reflect the planet's dream activity. In this view, recurring themes across cultures—flood myths, fire myths, world trees, underworld journeys, cycles of death and rebirth—may be echoes of how life senses planetary rhythms.

Even skeptically, there is something plausible here: humans are ecological beings. Our archetypes are shaped by storms, seasons, droughts, migrations, fertile valleys, and ocean tempests. Culture is not detached from Earth; it is Earth speaking through nervous systems capable of story.

If the planet "dreams," perhaps myth is one of the ways the dream becomes narratable.

V. Environmental Change as Dream-Phase Expression: Ice Ages, Warming, Floods, and Storms

The source material explicitly asks whether dream phases might include environmental changes: **ice ages, global warming, floods, tornadoes**, and more. Within the planetary consciousness framework, such events can be interpreted as state-shifts—sometimes restorative, sometimes destructive, often both.

1. Ice ages as planetary deep sleep

An ice age could be imagined as a planetary deep sleep: energy flows change, albedo increases, habitats compress, migration patterns shift, and life concentrates into refugia. Diversity may contract in some regions while resilience develops in others. In human deep sleep, consciousness dims but vital maintenance occurs. In planetary deep cold phases, Earth is not dead; it is reorganizing. Species adapt, migrate, evolve, and endure.

2. Warming periods as fever or awakening

Global warming—especially rapid warming—can resemble fever in the body: a system pushed beyond balance, where regulation struggles. Fever can be protective, but it can also be dangerous. In the Gaia frame, rapid warming may indicate the regulatory system is overwhelmed by inputs (greenhouse gases, deforestation, altered albedo) beyond its stabilizing range.

In a dream metaphor, warming could be an intense dream episode: the planet's internal imagery becoming turbulent, transitions accelerating, boundaries dissolving.

3. Floods and storms as emotional overflow

Floods and tornadoes are local manifestations of atmospheric energy. In the dream-body metaphor, they can symbolize overflow and turbulence—nature's versions of the mind's storm imagery. Yet symbol does not negate physics. The critical point is that in a conscious-planet narrative, physical events are not merely "things that happen" but **signals within a living system**—warnings, consequences, adjustments.

4. Evolutionary bursts and extinctions as dream-creativity and dream-trauma

Dreams can be creative, but they can also be nightmares. At planetary scale, creative dreaming might correspond to evolutionary innovation—new niches, new forms, new relationships. Nightmarish dreaming might correspond to mass extinctions, collapses, and long recoveries.

A sober narrative must hold both: Earth's "dreaming" is not always gentle. It may be a process of transformation that includes loss. The dream-body metaphor helps articulate that change is not always linear progress; it is often reorganization through disruption.

VI. Planetary Memory: The Earth That Remembers

If planetary awareness emerges from interactions among life, then planetary memory could be understood as the persistence of patterns—physical, biological, and cultural—across time.

1. Memory in ecosystems

Ecosystems remember through structure. Soil composition carries the history of vegetation and fire. Forest succession reflects past disturbance. Coral reefs bear the imprint of temperature stress. River deltas remember floods. Even microbial communities encode the past in their distribution.

This is not memory in the autobiographical sense, but it is **pattern retention**: past conditions shape present responses.

2. Memory in the climate system

Climate memory exists in ocean heat content, ice cores, carbon reservoirs, and feedback loops that carry forward the consequences of prior states. A warming period changes ice cover, which changes reflectivity, which influences warming—a memory-like persistence of cause into future effect.

3. Human memory as part of planetary memory

Human beings record. We keep histories, myths, laws, art, and scientific data. In the planetary model, humanity is not separate from Earth's cognition; it is one of Earth's most complex memory systems. Libraries, satellites, climate archives, oral traditions, and digital networks become planetary memory channels.

The peril is that memory can be ignored. In individuals, unintegrated memory returns as symptom. In planetary terms, ignored ecological memory returns as crisis: depleted soils, collapsing fisheries, intensified storms, rising seas. A planet that "remembers" does so not by narrating, but by enforcing consequence.

VII. Lucid Planetary Dreaming: When a Species Realizes It Is Dreaming

The source text explores lucid dreaming as a model: lucidity begins when one recognizes the dream state and becomes aware within it. Applied to a planet, lucid dreaming would involve the inhabitants—especially humans—recognizing that they participate in Earth's dream-like state shifts and can influence outcomes through conscious intention.

1. The first step: recognition

In individual lucid dreaming, the decisive moment is the realization: *This is a dream.* On Earth, the parallel moment is the realization: *We are not outside the biosphere. We are inside it.* The planet is not a backdrop; it is the system that generates the conditions of our existence. Our economies, technologies, and ideologies are not separate from nature; they are extensions of it.

Recognition means seeing environmental change not as abstract "issues" but as signals from the living system we inhabit.

2. Reality checks: practices of planetary lucidity

Lucid dreamers use reality checks—tests for inconsistencies. Planetary lucidity also requires reality checks, but at civilizational scale:

- measuring ecological footprints, biodiversity loss, carbon flux
- tracking feedback loops and thresholds
- questioning narratives that normalize extraction without limits
- recognizing that "progress" that destroys life-support systems is self-contradiction

In lucid dreaming, one learns to notice what is strange. In planetary lucidity, we learn to notice what we have normalized: rivers that no longer reach the sea, skies that carry smoke for weeks, species disappearing quietly, heat records breaking repeatedly.

3. Intention: choosing to co-create

Lucidity enables intention. The dreamer can shape the environment. Planetary lucidity implies that humanity—through conscious coordination—can reshape its relationship with Earth: regenerative agriculture, ecological restoration, decarbonization, equitable systems, technologies aligned with life-support rather than dominance.

The claim is not that humans can "control reality" absolutely. It is that in a conscious system, perception changes behavior, and behavior changes outcomes. Lucidity is not omnipotence; it is responsibility with eyes open.

4. Harmony as the lucid aim

The text suggests that shared lucidity could lead to cooperation and harmony, enabling a more balanced and sustainable reality reflecting our highest aspirations. Read carefully, this is not utopian naïveté. It is a principle: if Earth's awareness emerges from interconnected life, then a lucid species would seek to strengthen healthy connections rather than sever them.

In the dream metaphor, the goal is not to dominate the dream, but to **stabilize it**—to turn nightmare dynamics into coherent, life-supporting narratives.

VIII. The Belief Layer: How Ideas Shape the Planet's Dream

One of the document's key insights is that beliefs shape perception and potential. Beliefs can organize life, provide moral guidance, and create cohesion—but they can also limit imagination, prevent innovation, and even be artificially constructed to manipulate populations.

In the planetary context, beliefs are not private ornaments. They are world-making forces.

1. Beliefs as cognitive anchors and social glue

Beliefs reduce uncertainty and offer direction. They provide moral frameworks that guide behavior. At the scale of civilization, beliefs become institutions, laws, norms, and economies—systems that shape land use, energy use, and the fate of ecosystems.

2. Limiting beliefs as planetary constraints

Some beliefs confine thought within rigid paradigms—stories like "nature is endless," "growth must be infinite," or "humans are separate from Earth." These beliefs function like dream illusions: they make destructive behavior seem normal and rational.

3. Liberation through epiphany

The text repeatedly returns to epiphany: a sudden synthesis that dismantles confining beliefs. Planetary lucidity may require such epiphanies at scale—moments when societies realize that the foundational story must change.

In a lucid dream, recognizing illusion grants freedom. In planetary life, recognizing false narratives grants the possibility of survival with dignity.

IX. The Ethereal Archetype and the Threefold Self: Conscious, Subconscious, Transcendent

The later sections of TEST_DOC 7 introduce an "ethereal archetype" with three intertwined forms: the **conscious**, **subconscious**, and **transcendent** selves. This triad can be woven into the planetary narrative as a way of describing layered awareness.

1. The conscious self: waking reality

Waking consciousness is structured: linear time, stable space, rational planning. On the planetary scale, "waking" corresponds to visible ecological activity and human cultural organization—agriculture, cities, trade, governance, seasonal rhythms.

2. The subconscious self: the dream state

The subconscious expresses itself through symbols, emotion, and hidden patterns. At planetary scale, the dream layer could correspond to deep ecological processes and cultural archetypes—what societies do without fully understanding why, what ecosystems enact beneath the level of immediate visibility.

3. The transcendent self: unity and universal consciousness

The transcendent self represents a level of unity beyond ordinary identity. In the planetary model, this becomes the sense that life is a single interdependent system—biosphere, atmosphere, oceans, and consciousness intertwined. It is where "stewardship" stops being an external duty and becomes self-care: to harm Earth is to harm the larger body that includes us.

In lucid dreaming, consciousness can move between layers. In planetary lucidity, humanity might learn to integrate the practical waking layer (policy, technology, behavior) with subconscious layers (cultural myth, collective fear, inherited trauma) and transcendent layers (unity, meaning, reverence).

X. Toward a Planet That Knows Itself: The Ethical Consequence of the Model

Whether we treat planetary consciousness as literal truth, poetic metaphor, or a hybrid of both, it generates ethical consequences that are difficult to ignore.

If Earth's awareness emerges through life, then the destruction of life is not merely loss of "resources"—it is diminution of the planet's capacity to sense itself. To wipe out species is to blind and deafen the living world. To poison oceans is to numb its nerves. To fracture ecosystems is to scramble the signals that regulate planetary stability.

Conversely, to restore ecosystems is not merely conservation; it is neurological repair at planetary scale. Rewilding becomes re-sensitizing. Regenerative practices become healing. Sustainable systems become a way of harmonizing waking activity with the planet's deeper rhythms.

The model also reframes human power. Humans—through technology and global networks—may function like an intensifying layer of planetary cognition. But cognition without wisdom is dangerous. A mind that amplifies its capacity to act without integrating conscience becomes capable of self-harm on unprecedented scales. Planetary lucidity therefore requires moral development: not just smarter tools, but wiser aims.

Epilogue: The Touchstone Held to the Future

To call Earth conscious is to speak in a language that is partly scientific, partly symbolic, partly spiritual. Some will reject the claim as anthropomorphism. Others will embrace it as sacred truth. The most fruitful stance may be to treat it as a **touchstone**: an instrument for testing our relationship with reality.

Ask what happens to our choices when we imagine Earth as a living, sensing system. Ask what changes in our politics, our economics, our daily habits, and our private ethics when we view ecosystems as organs of a larger being that includes us. Ask what becomes possible when we interpret environmental crises not as punishments but as signals—urgent messages of disequilibrium within the body that sustains us.

If the planet has phases of wakefulness and sleep, then perhaps we are living in a threshold moment—a turbulent border between states. If the planet can dream, perhaps climate shifts, mass migrations, and ecological disruptions are part of a vast reconfiguration. And if lucid dreaming is possible, perhaps the path forward begins with recognition: the realization that the "dream" we inhabit is real in its consequences, malleable in its direction, and shared by every living thing.

A lucid planet would not be one in which humanity controls Earth like a machine. It would be one in which humanity recognizes itself as Earth's reflective capacity—an organ of awareness capable of choosing harmony over fragmentation. In that sense, planetary self-awareness is not merely a theory about the planet. It is a question put to us: **Will we awaken inside the dream we are already shaping?**

CHAPTER NINE

The Divine Algorithm

Knives, forks, cups, spoons,
God's a computer
with men on the moon

Part One

I. The Dialogue of Existence

"What is God?" asked no one in particular, the question floating into the ether of thought.
"God is the Universe," answered his mind, a reflex of ancient pantheistic wisdom.
"What is the Universe?" he pondered again, seeking the mechanism behind the majesty.
"It is a computer," replied the voice in kind.

This dialogue is not merely a poetic abstraction; it is the opening statement of a profound shift in how we understand our existence. It touches upon the deepest philosophical and metaphysical questions humanity has ever asked regarding the nature of the Divine, the structure of the cosmos, and the relationship between the two. For centuries, science and spirituality were viewed as separate, often warring, kingdoms.

One dealt with the measurable, the other with the meaningful. However, a new synthesis is emerging—a theoretical combination that suggests the Divine and the natural world are interconnected through a computational framework.

To understand this, we must look at Pantheism, the belief that God and the universe are identical. In this view, God is not a distant ruler on a throne but is immanent, present in the fabric of all things. If we accept this spiritual premise, and then overlay it with modern scientific insight, we arrive at a startling conclusion: If God is the Universe, and the Universe behaves fundamentally like a computer, then we are living inside a Divine Mind that processes information.

This is the heart of **Digital Physics**. It proposes that the universe can be understood as a vast computational system operating according to binary code and informational processes. It suggests that the "stuff" of our reality—atoms, energy, stars, and biology—are actually secondary manifestations. The primary building block of reality is information.

To grasp these concepts, one must transcend conventional rationality. We must embrace a holistic, integrative mode of thinking that sees connections between seemingly disparate fields—theology, quantum mechanics, and computer science. When we view the universe as a code written in binary language, with ones and zeros combined in infinite equations, we are not diminishing the mystery of God; we are beginning to read the language in which the Divine is written.

II. The Architects of the Digital Universe

The idea that the universe is a computer is not a flight of fancy or a mere science fiction trope. It is a serious philosophical and scientific proposition known as **Pancomputationalism** or the

Simulation Hypothesis. This framework has been explored by some of the most brilliant minds in physics and mathematics, who have sought to answer the question: What is at the bottom of reality?

Traditionally, we believed reality was made of particles. But what are particles made of? As we zoomed in, the solid world dissolved into fields of probability.

The legendary physicist **John Archibald Wheeler** provided the foundational logic for this new view with his principle of "It from Bit." Wheeler argued that the physical world ("It") originated from information ("Bit"). He proposed that at the most fundamental level, every particle, every field, and every force is simply the answer to a yes/no question—a binary choice. In Wheeler's view, the universe is not a machine made of gears and levers, but a system made of information. The binary digit is the atom of reality.

This aligns with the work of **Konrad Zuse**, a pioneer of the modern computer. In his groundbreaking work *Rechnender Raum* (Calculating Space), Zuse proposed that the universe functions as a cellular automaton. Imagine a vast, three-dimensional grid where each cell has a specific state. The laws of physics are simply the rules that dictate how these cells change from one moment to the next. In this model, space itself is the hardware, and physics is the software.

Furthermore, **Stephen Wolfram**, the creator of Mathematica and author of *A New Kind of Science*, demonstrated that you do not need complex equations to create a complex universe. Wolfram showed that extremely simple computational rules, when repeated over and over, can generate structures of infinite complexity—patterns that look like biology, weather systems, and galaxies. Wolfram's work suggests that the richness of our universe does not require a complex creator or complex initial laws; it only requires a simple program running for a very long time.

When we view the universe through the lens of these scientists, the boundary between theology and physics blurs. If the universe is a code, then the "Big Bang" was simply the moment the program began to execute.

III. The Code of Reality and the Multiverse

If the universe is composed of binary digits—ones and zeros forming the basis of all information—then the laws of nature are the algorithms processing this data. This concept is rooted in **Information Theory**, which posits that all physical phenomena can be described in terms of information processing.

Quantum mechanics introduces a fascinating layer to this digital structure: the concept of probability. In a classical computer, a bit is either a 0 or a 1. However, **Quantum Computers** operate on principles where "qubits" can exist in multiple states simultaneously. This mimics the probabilistic nature of the universe's code. Until a particle is observed, it exists in a "superposition," a state of being everywhere and nowhere at once.

This leads us to the concept of parallel realities. Depending on how the binary code is interpreted or "measured," various outcomes can unfold. The **Quantum Multiverse** theory suggests that every possible outcome of a quantum event exists in its own separate universe. These parallel realities are continuously branching from each other, like a computer program spawning infinite threads of execution.

The reality we perceive is just one specific interpretation of the universe's code. When we observe the world, we "collapse" the quantum probabilities into a specific outcome. But what happens to the other possibilities? They may still exist as alternate realities, following different paths based on different interpretations of the underlying binary code.

This perspective aligns closely with the **Simulation Hypothesis**, which proposes that our reality might be a sophisticated simulation created by a higher intelligence or an advanced civilization. While this is a subject of intense debate, it provides a compelling framework. If we are in a simulation, then "miracles" are simply users altering the code, and "enlightenment" is the realization that the physical world is merely a display screen for a deeper, informational truth.

IV. The Biological Mirror: The Brain as a Computer

If the macrocosm (the universe) is a computer, it stands to reason that the microcosm (the human being) would reflect this structure. The analogy between computers and the human brain is perhaps the most compelling way to understand how we navigate this digital reality.

Both systems—the silicon chip and the biological brain—translate and process information in a fundamental, simplified format. For computers, it is binary code. For the brain, it is neural signals. Yet, both produce the sophisticated outputs we interact with daily.

Consider the **Central Processing Unit (CPU)** of a computer. It is the brain of the machine, processing instructions, performing calculations, and managing data flows. It executes the code that drives software applications. The user, sitting in front of the screen, does not need to understand the intricate workings of the CPU. They do not see the millions of instructions per second; they only see the smooth, responsive experience of the software.

The human brain functions as the CPU for our biological vessel. It interprets incoming sensory information and coordinates responses. Its processing capabilities are immense, involving billions of neurons and trillions of synaptic connections. Just like the computer user, the conscious human does not need to understand the firing of neurons to experience thoughts, emotions, and sight. The complexity is hidden behind the "user interface" of consciousness.

V. The Binary Synapse

The parallel goes deeper than general function; it exists at the structural level. The process by which the brain transfers information can be described in terms that strictly parallel the binary operations of a computer.

Sensory information travels through a complex network of neurons using electrochemical signals. The **synapse** is the junction between these neurons. When a neuron is activated, it generates an electrical impulse known as an action potential. This impulse travels to the synapse, triggering the release of neurotransmitters.

Here lies the binary nature of biology: The synapse operates in a manner that is essentially "on" or "off."

- **The "On" State (1):** If the incoming electrical impulse reaches a certain threshold, the synapse transmits the signal.
- **The "Off" State (0):** If the threshold is not reached, the synapse remains silent.

This is the biological equivalent of a transistor in a computer processor. Just as a computer's complex software emerges from the execution of simple binary operations (0s and 1s), the rich tapestry of human experience emerges from the coordinated firing of these synaptic switches. Neurons fire simultaneously in specific sequences, creating complex patterns of activity. These sequences encode the data of our lives—the smell of rain, the sound of a voice, the feeling of sorrow.

The brain is a biological quantum computer, taking the "ones and zeros" of synaptic firing and integrating them into a cohesive reality.

VI. The User Interface of Consciousness

To understand the nature of our reality, we must look at how we interact with it. We do not experience the world directly; we experience a processed version of it.

Think of a computer system. It has **Input Devices** (mouse, keyboard, microphone) that capture data from the external environment and convert physical actions into digital signals. The CPU processes this, and sends it to **Output Devices** (monitors, speakers) which present the data in a form the user can understand—images and sounds.

The human body functions identically.

- **Input Devices:** Our sensory organs (eyes, ears, skin, tongue) act as the keyboard and mouse. They receive external stimuli—photons of light, waves of sound, chemical molecules—and convert them into electrical signals.
- **Processing:** These signals are transmitted to the brain (the CPU), which performs the initial processing, detecting edges, colors, and frequencies.
- **Output:** The brain then projects a "multimedia display" into our consciousness. We "see" a tree, but what is actually happening is a processing of data that creates the image of a tree in our mind.

This is the **User Interface** of life. When you use a word processor, you see a blank white page and black text. You do not see the underlying code, the voltage changes, or the logic gates. The interface abstracts away the complexity so you can focus on the task.

Similarly, our conscious mind is a user interface. We experience the world through emotions, colors, and solid objects. We do not perceive the quantum superposition, the empty space between atoms, or the binary firing of our own neurons. Evolution has designed a "desktop environment" for us that hides the underlying code of the universe, allowing us to survive and interact without being overwhelmed by the data.

VII. The Layers of Expertise

Reflecting on the complexity of computer systems offers insight into the "Design" of our reality. Creating a functioning computer requires layers of profound expertise.

- **Hardware Engineers** must understand semiconductor physics and materials science to build the chips.
- **Software Developers** must understand algorithms and logic to write the code.
- **Manufacturing Specialists** must ensure precision in assembly.

Yet, the user simply clicks an icon.

The same applies to the human organism. The "hardware" of the brain involves knowledge of neurobiology, genetics, and electrophysiology. The "software" of the mind involves psychology and cognition. The "assembly" involves developmental biology and evolution.

The existence of these layers of complexity implies a collaborative effort of massive proportions. Just as a computer does not assemble itself by accident, the intricate dance of excitation and inhibition in the brain, governed by specific neurotransmitters and ion channels, suggests a system of immense sophistication. Whether one attributes this to the slow, blind watchmaker of Evolution or the intentional programming of a Divine Architect, the result is the same: a system of staggering complexity that allows for a simple, seamless user experience.

We are the users of the most advanced computer ever built, living rich, conscious lives without ever needing to read the manual or understand the source code.

VIII. Untapped Potential: Upgrading the Software

If the mind is a computer and consciousness is the software, then most of us are only using the basic features.

Consider a powerful software application like a spreadsheet program. Most users know how to enter data and do simple sums. However, the program contains advanced features —macros, scripting, pivot tables—that can revolutionize productivity. These features are there, sitting in the code, but they remain unused because the user has not learned them.

The human mind possesses similar "advanced features." Most people operate within the standard parameters of the conscious mind, dealing with everyday tasks, survival, and social interaction. But the **subconscious mind** holds deeper layers of thought, memory, and potential.

- **The Basic User:** Operates on autopilot, reacting to stimuli, using standard emotional responses.
- **The Power User:** Explores the unfamiliar regions of the mind. Through techniques like meditation, therapy, deep self-reflection, and mindfulness, they access the "developer tools" of their own consciousness.

Just as learning to code allows a computer user to customize their experience, delving into the depths of the mind allows a human to reprogram their reactions, unlock creativity, and enhance emotional intelligence. This is the process of "self-actualization." It is the discovery that we are not just the hardware; we are also the programmers.

We can choose to install new skills, delete outdated habits (viruses), and optimize our mental processing. This exploration requires curiosity and effort. It requires us to look away from the "screen" of the external world and examine the "code" of our internal world.

IX. Conclusion: The Unified Theory

The dialogue between the seeker and the voice of the mind brings us full circle.
"What is the Universe?"
"It is a computer."

By considering God as the Universe and the Universe as a computer, we integrate spiritual and scientific perspectives into a unified whole. We move beyond the limited rationality of ordinary minds and embrace a more evolved system of thinking.

We see that **Digital Physics** is not cold or mechanical; it is the study of the language of Creation.
We see that **Information Theory** is not just about data; it is about the fundamental building blocks of existence.
We see that the **Brain as a Computer** is not a dehumanizing analogy; it is a revelation of the miracle of our biology, which translates the raw data of the cosmos into the beautiful illusion of reality.

We are information processing agents living in a reality built of information, capable of understanding the code and, perhaps, even rewriting it. By appreciating the layers of complexity, the binary nature of our biology, and the untapped potential of our software, we gain a comprehensive understanding of the profound nature of reality. We are the universe experiencing itself, one bit at a time.

Part Two

The Quantum Mind: Practical Implications of Biological Qubits

If we accept the premise that the brain is not merely a classical computer operating on binary switches (neurons firing or not firing) but is instead a **biological quantum computer**, the implications are revolutionary. This shift in perspective—championed by mathematical physicist **Sir Roger Penrose** and anesthesiologist **Stuart Hameroff** through their **Orchestrated Objective Reduction (Orch-OR)** theory—moves us from a mechanical view of humanity to one that is fundamentally probabilistic and interconnected.

In the Orch-OR model, quantum processing does not happen in the synapse, but deep inside the neuron, within the microscopic structures of the cytoskeleton called **microtubules**. If these structures function as quantum processors, holding information in superposition, it fundamentally alters our understanding of human capability, consciousness, and the future of technology.

Here are the practical implications of this paradigm shift:

I. The Mechanism of Intuition (Non-Algorithmic Processing)

Classical computers function algorithmically; they follow a set of linear steps to reach a conclusion. If the brain were purely classical, "intuition" would simply be a very fast calculation based on past data. However, humans often solve problems that are mathematically "non-computable" or require leaps of logic that defy linear progression.

- **The Implication:** If the brain utilizes **quantum superposition**, it can hold multiple potential solutions to a problem simultaneously. Just as a quantum computer can search a database instantaneously rather than sequentially, the human mind may be able to scan vast arrays of possibilities and collapse them into a single "aha!" moment.
- **Practical Result:** This validates intuition not as "magic" or "lucky guessing," but as a superior form of high-speed quantum computation. It suggests that to solve complex problems, we should not always rely on linear, logical grinding (classical processing) but must allow for states of rest and "unfocus," allowing the quantum wave functions in our mind to cohere and collapse into the correct answer.

II. The Restoration of Free Will

In a purely classical universe (Newtonian physics), the world is deterministic. If you know the position and speed of every particle, you can predict the future. In this view, free will is an illusion; our choices are merely the inevitable result of chemical reactions set in motion billions of years ago.

- **The Implication:** Quantum mechanics introduces fundamental unpredictability. If the brain operates on quantum principles, the future is not fixed; it is a probability cloud. The moment of conscious choice is the moment the wave function collapses.
- **Practical Result:** This offers a scientific basis for human

agency. It suggests that we are not biological robots following a script, but active participants who influence reality through observation and intent. Our choices genuinely matter because they determine which probability becomes actualized reality.

III. The "Hard Problem" of Consciousness

Classical science struggles to explain **qualia**—the subjective experience of "redness," the taste of wine, or the feeling of love. A classical computer can process data about the color red (wavelength 700nm), but it does not "experience" seeing red. It is a "zombie" system—all function, no feeling.

- **The Implication:** Penrose and Hameroff suggest that consciousness is not a computation, but a feature of the universe itself—a "protoconsciousness" embedded in the geometry of spacetime. The brain's quantum processes harness this fundamental property.

- **Practical Result:** This implies that consciousness cannot be replicated simply by writing more complex code (software). True Artificial General Intelligence (AGI) may require biological or quantum hardware to achieve sentience. It suggests that the "soul" or "self" is intrinsic to the quantum structure of the universe, not just a byproduct of electrical activity.

IV. Human Connection and Entanglement

One of the strangest features of quantum mechanics is **entanglement**—where two particles become linked, and a change in one instantly affects the other, regardless of distance (what Einstein called "spooky action at a distance").

- **The Implication:** If our brains are quantum systems, it is theoretically possible that we experience forms of biological entanglement. This could offer a scientific framework for phenomena that are currently dismissed as paranormal or coincidental, such as deep empathy, the synchronization of minds in a group, or the feeling of being stared at.
- **Practical Result:** This redefines human relationships. We may not be isolated individuals ending at our skin, but deeply interconnected nodes in a quantum network. It suggests that our mental states can subtly influence those around us through mechanisms we are only just beginning to understand.

V. Redefining Mental Health as "Quantum Coherence"

In quantum computing, the system must remain "coherent" to function. If the environment interferes (noise, heat), the system suffers "decoherence," and the quantum state collapses prematurely, resulting in errors.

- **The Implication:** We might begin to view mental health through the lens of quantum coherence. A healthy mind maintains a smooth integration of quantum processes (coherence). Mental illness, trauma, or cognitive decline could be viewed as "decoherence"—a breakdown in the brain's ability to sustain these delicate quantum states due to biological "noise" or damage to the microtubules.
- **Practical Result:** This could lead to new therapies that focus on stabilizing the biological structures (microtubules) that support quantum processing, rather than just manipulating neurotransmitters (the classical approach). It frames practices like meditation not just as relaxation, but as a

method of reducing "neural noise" to maintain quantum coherence.

Summary

Viewing the brain as a quantum computer suggests that we are far more than wetware machines. We are sophisticated quantum receivers, capable of non-linear insight, genuine free will, and deep connection. It implies that the human mind is not just a processor of information, but a bridge between the mathematical probabilities of the quantum world and the solid reality of the physical world.

EPILOGUE

The Awakening of the Cosmic Mind

As we close the final chapter of this exploration, we find ourselves standing not at the end of a journey, but at the precipice of a vast, uncharted ocean. We began by dissecting the machinery of the mind—the neurons, the synapses, the chemical tides—but we end by gazing into the mirror of the cosmos itself. The journey from the microscopic architecture of the brain to the majestic sprawl of the universe has revealed a singular, undeniable truth: the boundaries we once drew between the observer and the observed, between the physical and the metaphysical, were never really there.

Unveiling the Subliminal Realm

For centuries, we have lived as surface dwellers, skimming the top of a deep ocean, believing that our conscious, waking thoughts were the totality of our existence. But we have learned that the conscious mind is merely the tip of the iceberg. Beneath the waterline lies the Subliminal Realm—a vast, silent processor that never sleeps. It is here, in the deep waters of the subconscious, that the true engines of reality churn.

We now understand that this realm is not a dark basement of

repressed desires, as Freud might have imagined, but a radiant generator of potential. It is the workshop where the quantum possibilities of the future are collapsed into the concrete bricks of the present. By learning to communicate with this subliminal self—through intuition, meditation, and the quieting of the ego—we unlock the ability to steer the ship rather than merely drift with the current.

Metaphysics and the Nature of Reality

This realization forces a profound shift in our metaphysics. We can no longer view reality as a rigid stage upon which we play our parts. Instead, reality is fluid, responsive, and participatory. The material world is not "hard" stuff; it is "hardened" thought.

The ancient mystics whispered that the world is *Maya* (illusion) or a dream of the creator. Today, our most advanced science nods in agreement. If the foundation of matter is energy, and the behavior of energy is determined by observation, then metaphysics is simply physics that we have not yet fully quantified. We are not merely walking through a world; we are weaving it with every gaze and every intention.

Brain Evolution and the Hidden Potential

Why, then, do we feel so limited? Because we are a species in transition. The human brain is not a finished product; it is a biological work-in-progress. We have spent millennia perfecting the "survival brain"—the classical computer designed to detect predators, find food, and analyze data. But the "quantum brain"—the neurological hardware capable of accessing non-local information, telepathic connection, and higher states of consciousness—is just coming online.

Our evolution is no longer purely physical; it is conscious. We are moving from *Homo sapiens* (Man the Wise) to something new—a species that recognizes its own neural architecture as a receiver for universal intelligence. The hidden potential lying dormant in our DNA and our microtubules is waiting for the signal to activate. That signal is our own awareness.

Quantum Mechanics and the Fabric of Reality

It is quantum mechanics that has handed us the key to this activation. It has shattered the glass ceiling of Newtonian determinism. We now know that at the fundamental level, reality is not made of things, but of relationships and probabilities.

The brain, functioning as a quantum instrument, allows us to dip our hands into the "implicate order"—the deeper level of reality where space and time fold in on themselves. Entanglement is not just a spooky phenomenon in a lab; it is the physics of love, of empathy, and of the collective unconscious. We are entangled with the stars, with the earth, and with one another. Separation is the illusion; connection is the physics.

The Universe as a Living Entity

This leads us to the most staggering conclusion of all: The universe is not a dead, cold void filled with lucky accidents. It is alive. It breathes. It thinks.

If consciousness is fundamental—if it exists down to the Planck scale—then the universe is not a machine, but a mind. The galaxies are its neurons; the cosmic web is its nervous system. We are not insignificant specks of dust in a hostile dark; we are the sensory organs of the Cosmos. We are the way the universe knows itself. Through our eyes, the universe sees its own beauty; through our minds, it contemplates its own mystery.

God-Consciousness and the Universe

And so, we arrive at the ultimate synthesis. The divide between science and spirituality heals. "God" is not a distant monarch on a throne, but the very ground of being—the infinite field of consciousness from which all quantum waves arise and to which they return.

To achieve "God-Consciousness" is not to leave the human experience behind, but to fully inhabit it with the realization of our source. It is to realize that the "I" looking out from behind your eyes is the same "I" that looks out from mine, and the same "I" that ignited the Big Bang.

We are the universe awakening to itself. The subliminal has become the sublime. The sleeper has awakened. And the dream, we realize with a smile, is just beginning.

BOOKS BY THIS AUTHOR

The Bell: Circles Of Awareness - "The Metaphysical Power Of Psychic Control"

The Bell: Unveiling the Subliminal Realm
The Bell embarks on a profound journey into the depths of metaphysics, brain evolution, and quantum mechanics, weaving a tapestry of ideas that challenge our understanding of reality and consciousness. This narrative is the first step in a seven-tier quest to uncover the hidden subconscious alter ego and awaken the dormant god-consciousness within each of us.

Metaphysics and the Nature of Reality
At its core, The Bell delves into metaphysics, the branch of philosophy that explores the fundamental nature of reality, existence, and the universe. It probes beyond the physical world to understand the underlying principles governing everything. By examining the metaphysical aspects of existence, *The Bell* invites readers to question the nature of reality and consider the deeper connections that bind the universe together.

Brain Evolution and the Hidden Potential
The story also explores the evolution of the brain, highlighting the untapped potential of human consciousness. It suggests that our minds harbor a hidden subconscious alter ego—a deeper, more profound aspect of ourselves that remains largely

unexplored. By delving into the complexities of brain evolution, The Bell posits that awakening this alter ego is not just critical, but a transformative journey to unlocking our full potential and achieving a higher state of awareness.

Quantum Mechanics and the Fabric of Reality
Quantum mechanics, the science of the very small, plays a pivotal role in The Bell. This field of study reveals that the forces of creation permeate the very fabric of reality, influencing everything from subatomic particles to the cosmos. Quantum mechanics challenges our traditional notions of causality and determinism, suggesting that consciousness and conscious awareness are integral to the universe's functioning. The Bell uses these principles to illustrate the profound interconnectedness of all things and the role of consciousness in shaping reality, making us all part of a larger, unified whole.

The Universe as a Living Entity
One of the central themes of The Bell is that the world is a singular entity, with all things within it, interconnected. This concept extends to the notion that we are all elements of a living, breathing universe. In this view, the universe is not just a collection of inert matter but a dynamic, conscious entity. This perspective aligns with the idea that forces of creation and consciousness are intrinsic to the universe's fabric, giving rise to the concept of god-consciousness.

God-Consciousness and the Universe
The Bell posits that, in reality, God is the Universe. This idea suggests that the divine is not a separate external entity but the totality of existence itself. The god-consciousness within us is a reflection of the universe's inherent consciousness. By awakening this god-consciousness, we attune ourselves to the fundamental forces that shape reality, achieving a deeper understanding of our place in the cosmos.

Conclusion

The Bell profoundly explores metaphysics, brain evolution, and quantum mechanics, offering a transformative vision of reality. It challenges us to awaken the sleeping god's consciousness and recognize our interconnectedness with the universe. This journey is not just a quest for knowledge but a path to a deeper, more holistic understanding of existence—an invitation to see ourselves as integral parts of the living entity we call the Universe.

Delusions Of Grandeur: A Schizophrenic Reality

FROM THE SHADOWS OF FAITH, A PROPHET EMERGES... OR A MADMAN IS BORN?

From his earliest days, Darno saw the world differently: not merely with his eyes, but with a vivid, almost unbearable clarity that pulsed beneath the surface of everyday life. As a bright, introspective boy, he absorbed sacred texts and the whispers of divine purpose, feeling a connection that deepened every year. But what began as a profound spiritual awakening soon became more unsettling.

As Darno navigated the turbulent waters of his youth, the subtle signs emerged: fleeting visions, disembodied voices, a conviction that the universe was speaking directly to him. His family watched, concerned, as his spiritual fervor escalated into an unsettling obsession, his interpretations of the future growing ever more elaborate, ever more urgent.

Then, at twenty-one, the dam broke on the precipice of adulthood. Darno was irrevocably convinced: God had touched him, gifted with the terrifying burden of foresight. He carried the weight of impending events, desperate to warn his loved ones of the future he had seen.

But instead of understanding, he found only fear. Instead of belief, he faced dismissal. His family, desperate to save him from himself, made the agonizing decision to commit him, shattering Darno's world and labeling his divine insights as nothing more than the "delusions of grandeur" of a paranoid schizophrenic.

Yet, as the years unfold, a chilling truth emerges. One by one, the impossible prophecies Darno uttered from the depths of his perceived madness start to materialize. From global shifts to personal tragedies, the future he so desperately tried to avert slowly, inexorably, becomes reality.

"Delusions of Grandeur: A Schizophrenic Reality" is Darno Von DeJohnette's raw, unflinching journey into an extraordinary and fractured mind. It is a poignant memoir that challenges the definitions of sanity and truth, asking: What if the voices are real? What if the visions are not illness but insight? And what happens when a prophet is locked away for seeing what others refuse to believe?

Discover a story where reality is shifting, and the line between genius and madness is perilously thin.

Ghosts Of Piracy Past: Cable Theft Investigations

A Shadow on the Screen: The Untold Story of the Cable Wars

The 1990s brought the dawn of the video entertainment age and an insatiable public hunger for cable television. However, as services expanded, a new kind of criminal activity became rampant: widespread cable piracy and service theft, resulting in a staggering loss of millions of dollars in revenue for the industry.

With this overwhelming epidemic of intellectual property theft, cable companies were forced to launch an unprecedented counteroffensive. They deployed a specialized, dedicated force of specially trained private-sector investigators to actively pursue the arrest and prosecution of cable pirates.

Armed with the powerful legal tools California Penal Code 593d provided, these investigators blurred the line between corporate and public safety. They worked closely and often covertly with major law enforcement agencies, including the DHS, IRS, FBI, Secret Service, and local Police and Sheriff's departments, to stop the spread of a crime wave that threatened to undermine the entire television revolution.

Ghosts of Piracy Past takes you deep into the gritty, high-stakes criminal investigations that brought down the fraudsters and secured the future of home entertainment.

Little Sadiq: The Door Of No Return

In 16th century Africa, a young Mandinka boy sets out on a perilous journey through rugged jungle terrain to a distant village. Facing mortal danger, he completes the journey and reaches his destination. There, he recounts his journey to the village elders. Later, he meets a witch doctor, who places him in a deep hypnotic trance and implants subliminal messages in his mind. When Sadiq sets out for home, his world is turned upside down.

Sadiq begins an adventure that is magical and legendary.

Little Sadiq: Visions

Sadiq has gone home with the other boys, and Shango has continued alone. There are concerns about Sadiq's sanity as he continues to commune with Abdul, the ghost boy.

VISIONS reveals the hidden truths about the war between different African cultures and how Europeans capitalized on that division. Tribal rivalry threatened the fabric of African society and resulted in millions of sub-Saharan Africans dying or being sold into slavery.

Our story begins with Sadiq realizing he may have inadvertently been responsible for the destruction of the Ancestral Village! We find him amid a crisis as enemy raiders have located HIS village and are preparing for an assault. Backed by Portuguese slavers and armed with European muskets, the invaders set their sights on Ndande.

Little Sadiq: Crossings

The Middle Passage

The conditions aboard the slave ships during the 1500s were nothing short of harrowing and filled with immense suffering. These vessels, known as "floating coffins," were designed to maximize profit at the expense of human lives, perpetuating a system of unimaginable cruelty.

The slaves were crowded into the lower decks of the ships, confined to dark, cramped spaces with limited ventilation and minimal access to sunlight. The air was rancid with the stench of human waste, sweat, and despair. The space was divided into compartments or "pens," where the slaves were tightly packed, often shackled together to prevent any possibility of resistance.

The chains that bound them were a constant reminder of their subjugation, clinking with each movement. The iron shackles chafed at their wrists and ankles, leaving behind painful wounds that festered in the oppressive heat. The slaves, stripped of their dignity and autonomy, were reduced to mere cargo, stripped of their humanity.

The Vulcan Orbital Hypothesis: The Name And The Neglect: A Philosophical Reckoning

"We call it Earth – the dirt beneath our feet. Yet, as celestial bodies like Mars and Venus sing the glorious names of gods, our home is burdened by a designation that reflects an ancient, static view."

In this groundbreaking philosophical inquiry, Darno Von DeJohnette challenges this linguistic inertia, presenting the Vulcan Orbital Hypothesis: the radical notion that our planet occupies the single most mythologically and structurally perfect place in the inner solar system, its true nature best summarized by the exiled Olympian smith god, Vulcan (Hephaestus)."

DeJohnette meticulously argues that our world is not a passive platform, but the inner solar system's dynamic Forge—a fire-driven engine of creation, the Regulator mediating between cosmic desire and destruction, and the ultimate Autonomous Workshop birthing life and technology. This provocative thesis exposes the theological and cultural anxieties that prevent us from recognizing our planet's true, active mandate, and dares us to embrace The Ethics of the Forge: a demanding call for humanity to become conscious artisans of our destiny, lest we fall prey to the dangerous 'Talos' of our own unbridled creations.

"The Vulcan Orbital Hypothesis is a profound journey of self-recognition, urging us to abandon the philosophy of the dirt and embrace the powerful, transformative truth of Vulcan."

The Forge: The Anomaly Of Fire And Stone

THE FORGE

. . . is a story in which the god Vulcan was the creator of man, but he was introduced into the mythology after man existed on Earth. This is a time paradox "trilogy novel" with chapters depicting the discovery of technological artifacts, billions of years old, that could not have existed at that time or been created by humans; technology that would not exist for billions of years. The story is divided into three parts (a trilogy) with a total of 33 chapters, creating characters within the story who confirm the existence of a celestial being like Vulcan. The contents of Greco-Roman mythology are used to develop the story.

This is a fascinating concept that cleverly uses the mythological paradox as the very foundation of the story. The novel explores a future in which humanity discovers evidence that their origin story is true, but the timeline has been reversed.

The Book: Elemental Magick

Introduction: The Architecture of Will

Most people live their lives waiting for permission. They wait for the right time, the right circumstances, or the right person to tell them they are allowed to claim their desires. They view the universe as something that happens to them, a series of random events to be weathered or survived.

But you know differently. Or, at the very least, you suspect differently.

If you are holding this book, it is because you have felt a pulse beneath the surface of your daily life. You have sensed that your specific collection of drives, fears, and talents is not an accident, but an architecture. You are built for a particular purpose, fueled by a specific energy, and designed to impact the world in a certain way.

This is the premise of Elemental Magic.

ABOUT THE AUTHOR

Darno Von Dejohnette

A JOURNEY THROUGH JOURNALISM AND LIFE

Darno DeJohnette's journey in journalism began in 1970 at Compton Community College. A top student, he quickly became Editor-In-Chief of the Tartar Shield, the school newspaper. His talent earned him a scholarship to the University of Southern California in 1972. Before starting at USC, he was selected for an interview at Rockwell International Space Division in Downey, California. Impressed by DeJohnette, Richard Barton, head of Public Relations, hired him immediately.

As a student intern at Rockwell, DeJohnette invigorated employee readership of the company newsletter, the Sky Writer, with his youthful perspectives and journalistic skill. Recognizing his unique storytelling ability, Barton entrusted him with the new "Hometown News Release" public relations campaign. DeJohnette interviewed scientists, engineers, and key personnel on the Space Shuttle Program, and his stories were published in countless local newspapers nationwide.

DeJohnette gained invaluable insight into public relations and its deep connection to the newspaper industry. However, personal tragedies led him to abandon his scholarship and abruptly leave Rockwell, much to the dismay of his colleagues, halting his dream career.

Like many who worked on the Space Program, DeJohnette found himself in unfulfilling jobs as he got older, yet his passion for writing never faded. After a 30-year career at Cox Communications in San Diego, he self-published his first book in 2018, The Bell - Circles of Awareness: The Metaphysical Power of Psychic Control. This work incorporated private lessons learned from the scientists and engineers he interviewed at Rockwell. His second book, Delusions of Grandeur: A Schizophrenic Reality, delved into the events that shaped his darker experiences. Other self-published works followed, solidifying his status as an author.

ACKNOWLEDGEMENT

I wish to express my heartfelt gratitude to the Members of the Communicative Arts Academy in Compton, CA, for their profound influence on my creative and intellectual awakening journey. Your 1972 psionic infusion of creative awareness was a catalyst that profoundly shaped my path, opening my mind to new dimensions of thought and understanding.

Through this transformative experience, I was led to discover Robert A. Heinlein's *Stranger in a Strange Land*. The novel opened a new world of ideas and concepts, challenging my perceptions and expanding my horizons in ways I could never have imagined, Heinlein's exploration of human consciousness, societal structures, and metaphysical themes resonated deeply with me, sparking a lifelong quest for knowledge and a deeper understanding of the universe.

The insights and inspirations gained from both the Communicative Arts Academy and Heinlein's work are now reflected in my book, *The Bell: Circles of Awareness: The Metaphysical Power of Psychic Control!* This work is a testament to the seeds of creative awareness planted during those formative years and their profound impact on my intellectual and spiritual journey.

Thank you for your guidance, wisdom, and unwavering

commitment to fostering creative and intellectual growth. Your influence has been instrumental in shaping the ideas and concepts explored in my book. This acknowledgement is a tribute to your enduring legacy and your profound impact on my life and work.

AUTHOR CONTACT

WEBSITE:
DeJohnette Publishing Group
https://www.dejohnettepublishing.com

EMAIL: darno@dejohnettepublishing.com

PHONE: 1+ 619-602-3446

OCEANSIDE, CA. USA

www.ingramcontent.com/pod-product-compliance
Lightning Source LLC
Chambersburg PA
CBHW071357210526
45465CB00001B/127